London Mathematical Society Student Texts 40

Dynamical Systems and Ergodic Theory

Mark Pollicott
University of Manchester

Michiko Yuri
University of Sapporo

CAMBRIDGE
UNIVERSITY PRESS

PUBLISHED BY THE PRESS SYNDICATE OF THE UNIVERSITY OF CAMBRIDGE
The Pitt Building, Trumpington Street, Cambridge CB2 1RP, United Kingdom

CAMBRIDGE UNIVERSITY PRESS
The Edinburgh Building, Cambridge, CB2 2RU, United Kingdom
40 West 20th Street, New York, NY 10011-4211, USA
10 Stamford Road, Oakleigh, Melbourne 3166, Australia

First published 1998

Printed in the United Kingdom at the University Press, Cambridge

A catalogue record for this book is available from the British Library

Library of Congress Cataloguing in Publication data

Pollicott, Mark.
Dynamical systems and ergodic theory / Mark Pollicott, Michiko Yuri.
    p.   cm. – (London Mathematical Society student texts ; 40)
Includes bibliographic references (p.   –) and index.
ISBN 0 521 57294 0 – ISBN 0 521 57599 0 (pbk.)
1. Topoligical dynamics. 2. Ergodic theory. I. Yuri, Michiko,
1956– . II. Title. III. Series.
QA611.5.P65   1998
515'.42–dc21   97-8812   CIP

ISBN   0 521 57294 0   hardback
ISBN   0 521 57599 0   paperback

CONTENTS

INTRODUCTION

This book is intended as an introduction to both dynamical systems and ergodic theory. Our aim is to give a direct and detailed introduction to the basic theory, suitable as a text for advanced undergraduate students or beginning graduate students in mathematics.

The notes divide naturally into three parts. The first part (chapters 1-6) concentrates on topological dynamics. The second part (chapters 7-12) deals with ergodic theory and measurable dynamics. The third part (chapters 13-16) consists of more advanced material to supplement the two earlier parts.

Each of the first two parts is intended to be essentially self-contained, as is illustrated by the following diagram of the relationships between chapters:

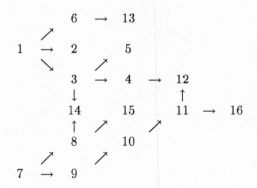

The areas of dynamical systems and ergodic theory are rich in connections with other subjects (e.g. number theory, geometry, statistics, mathematical physics, biology, etc.). In the course of these notes we have tried to motivate the general theory with some important applications (particularly to number theory, in chapter 2 and chapter 16).

There are already a number of excellent books on dynamical systems and ergodic theory (e.g. Devaney's *Introduction to Chaotic Dynamical Systems*) Walters' *Introduction to Ergodic Theory* and *Introduction to the Modern Theory of Dynamical Systems* by Katok and Hasselblatt).

To pre-empt any comparison with these fine texts, we should emphasize that this book is intended to be a more modest introduction to the subject. The reader who would like to find out more about dynamical systems and ergodic theory will find much more in these books.

Mark Pollicott
Department of Mathematics
Manchester University

Michiko Yuri
Department of Business Administration
Sapporo University

PRELIMINARIES

1. Conventions. The book is divided into 16 chapters, each subdivided into sections numbered in order (e.g. chapter 12, section 3 is numbered **12.3**). Within each chapter results (Theorems, Propositions or Lemmas) are labelled by the chapter and then the order of occurrence (e.g. the fifth result in chapter 3 is `Proposition` 3.5). The exceptions to this rule are: sublemmas which are presented within the context of the proof of a more important result (e.g. the proof of Theorem 2.2 contains Sublemmas 2.2.1 and 2.2.2); and corollaries (the corollary to Theorem 5.5 is Corollary 5.5.1).

We denote the end of a proof by ■.

Finally, equations are numbered by the chapter and their order of occurrence (e.g. the fourth equation in chapter 5 is labelled (5.4))

2. Notation. We shall use the standard notation: $\mathbb{R}$ to denote the *real numbers*; $\mathbb{Q}$ to denote the *rational numbers*; $\mathbb{Z}$ to denote the *integer numbers*; $\mathbb{N}$ to denote the *natural numbers*; and $\mathbb{Z}^+$ to denote the non-negative integers. We use the convenient convention that: $\mathbb{R}/\mathbb{Z} = \{x + \mathbb{Z} : x \in \mathbb{R}\}$ (which is homeomorphic to the standard unit circle); $\mathbb{R}^2/\mathbb{Z}^2 = \{(x_1, x_2) + \mathbb{Z}^2 : (x_1, x_2) \in \mathbb{R}^2\}$ (which is homeomorphic to the standard 2-torus); etc. However, for $x \in \mathbb{R}$ we denote the corresponding element in $\mathbb{R}/\mathbb{Z}$ by $x \pmod 1$ (and similarly for $\mathbb{R}^2/\mathbb{Z}^2$, etc.).

We denote the interior of a subset A of a metric space by $\text{int}(A)$, and we denote its closure by $\text{cl}(A)$.

If $T : X \to X$ denotes a continuous map on a compact metric space then T^n $(n \geq 1)$ denotes the composition with itself n times.

If $T : I \to I$ is a C^1 map on the unit interval $I = [0, 1]$ then T' denotes its derivative.

3. Prerequisites in point set topology (chapters 1-6). The first six chapters consist of various results in topological dynamics for which the only prerequisite is a working knowledge of point set topology for metric spaces. For example:

THEOREM A (BAIRE). *Let X be a compact metric space; then if $\{U_n\}_{n \in \mathbb{N}}$ is a countable family of open dense sets then $\bigcap_{n \in \mathbb{N}} U_n \subset X$ is dense.*

THEOREM B (SEQUENTIAL COMPACTNESS). *Let X be a metric space; then X is compact if and only if every sequence $(x_n)_{n\in\mathbb{N}}$ in X contains a convergent subsequence.*

THEOREM C (ZORN'S LEMMA). *Let Z be a set with a partial ordering. If every totally ordered chain has a lower bound in Z then there is a minimal element in Z.*

Two good references for this material are [4] and [5]

4. Pre-requisites in measure theory (chapters 7-12). Chapters 7-12 form an introduction to ergodic theory, and suppose some familiarity (if not expertise) with abstract measure theory and harmonic analysis. The following results will be required.

THEOREM D (RIESZ REPRESENTATION). *There is a bijection between*

(1) *probability measures μ on a compact metric space X (with the Borel sigma algebra),*
(2) *Continuous linear functionals $c : C^0(X) \to \mathbb{R}$,*

given by $c(f) = \int f d\mu$.

THEOREM E. *Let $(X, \mathcal{B}, \mu)$ be a measure space. For every linear functional $\alpha : L^1(X, \mathcal{B}, \mu) \to L^1(X, \mathcal{B}, \mu)$ there exists $k \in L^\infty(X, \mathcal{B}, \mu)$ such that $\alpha(f) = \int f \cdot k d\mu$, $\forall f \in L^1(X, \mathcal{B}, \mu)$* [3, p.121].

In proving invariance of measures in examples the following basic result will sometimes be assumed.

THEOREM F (KOLMOGOROV EXTENSION). *Let $\mathcal{B}$ be the Borel sigma-algebra for a compact metric space X. If μ_1 and μ_2 are two measures for the Borel sigma-algebra which agree on the open sets of X then $m_1 = m_2$* [3, p. 310].

The following terminology will be used in the chapter on ergodic measures. Given two probability measures μ, ν we say that μ is *absolutely continuous* with respect to ν if for every set $B \in \mathcal{B}$ for which $\nu(B) = 0$ we have that $\mu(B) = 0$. We write $\mu << \nu$ and then we have the following result.

THEOREM G (RADON-NIKODYM). *If μ is absolutely continuous with respect to μ then there exists a (unique) function $f \in L^1(X, \mathcal{B}, d\nu)$ such that for any $A \in \mathcal{B}$ we can write $\mu(A) = \int_A f d\nu$.*

We usually write $f = \frac{d\mu}{d\nu}$ and call this the *Radon-Nikodym derivative* of μ with respect to ν.

We call two measures μ, ν *mutually singular* if there exists a set $B \in \mathcal{B}$ such that $\mu(A) = 0$ and $\nu(A) = 1$. We then write $\mu \perp \nu$.

In chapter 8 we shall need a passing reference to Lebesgue spaces. A *Lebesgue space* is a measure space which is measurably equivalent to the

the union of unit intervals (with the usual Lebesgue measure) with at most countably many points (with non-zero measure).

In chapter 11 we shall use the following result.

THEOREM H (DOMINATED CONVERGENCE). *Let $h \in L^1(X, \mathcal{B}, \mu)$ and let $(f_n)_{n \in \mathbb{Z}^+}$ $\subset L^1(X, \mathcal{B}, \mu)$, with $|f_n(x)| \leq h(x)$, converge (almost everywhere) to $f(x)$; then $\int f_n d\mu \to \int f d\mu$ as $n \to +\infty$.*

Good general references for this material are [1], [2], [3].

5. Subadditive sequences. A simple result which proves its worth several times in these notes is the following.

THEOREM F (SUBADDITIVE SEQUENCES). *Let $(a_n)_{n \in \mathbb{N}}$ be a sequence of real numbers such that $a_{n+m} \leq a_n + a_m$, $\forall n, m \in \mathbb{N}$ (i.e. a subadditive sequence); then $a_n \to a$, as $n \to +\infty$, where $a = \inf\{a_n : n \geq 1\}$*

PROOF. First note that $a_n \leq a_1 + a_{n-1} \leq \ldots \leq na_1$, and so $a \leq a_1$ For $\epsilon > 0$ we choose $N > 0$ with $a_N < N(a + \epsilon)$. For any $n \geq 1$ we can write $n = kN + r$, where $k \geq 0$ and $1 \leq r \leq N - 1$. Then

$$a_n \leq a_{kN} + a_r \leq ka_N + a_r \leq ka_N + \sup_{1 \leq r \leq N} a_r$$

and we see that

$$\limsup_{n \to +\infty} \frac{a_n}{n} \leq \limsup_{k \to +\infty} \frac{ka_N + \sup_{1 \leq r \leq N} a_r}{kN} = \frac{a_N}{N} \leq a + \epsilon.$$

This shows that $\frac{a_n}{n} \to a$, as required.

■

References

1. P. Halmos, *Measure Theory*, Van Nostrand, Princeton N.J., 1950.
2. K. Partasarathy, *An Introduction to Probability and Measure Theory*, Macmillan, New Delhi, 1977.
3. H. Roydon, *Real Analysis*, Macmillan, New York, 1968.
4. G. Simmons, *Introduction to Topology and Modern Analysis*, McGraw-Hill, New York, 1963.
5. W. Sutherland, *Introduction to Topological and Metric spaces*, Clarendon Press, Oxford, 1975.

CHAPTER 1

EXAMPLES AND BASIC PROPERTIES

In this chapter we shall introduce some of the basic dynamical properties associated to continuous maps $T : X \to X$ on compact metric spaces.

1.1 Examples

To set the stage, we begin with some standard examples of continuous maps (transformations) which will be used to illustrate different properties.

EXAMPLE 1 (DOUBLING MAP). Let X denote the unit interval with its endpoints identified, $X = \mathbb{R}/\mathbb{Z}$. Define a continuous map $T : X \to X$ by $T(x) = 2x$ (mod 1), i.e.

$$Tx = \begin{cases} 2x & \text{if } 0 \leq x < \frac{1}{2}, \\ 2x - 1 & \text{if } \frac{1}{2} \leq x \leq 1. \end{cases}$$

This is usually called the "doubling map", since it doubles distances on X. An equivalent formulation would be if we let $X = K := \{z \in \mathbb{C} : |z| = 1\}$ and then define $T : X \to X$ by $T(e^{2\pi i \theta}) = e^{2\pi i 2\theta}$, where $0 ¡ \theta < 1$. This is equivalent in the sense that there is a homeomorphism $\rho : \mathbb{R}/\mathbb{Z} \to K$ given by $\rho(\theta + \mathbb{Z}) = e^{2\pi i \theta}$ which relates the two transformations. We shall return to this notion of equivalence (or conjugacy) in chapter 3.

EXAMPLE 2 (ROTATIONS ON THE CIRCLE). Let $X = \mathbb{R}/\mathbb{Z}$ and fix a number $\alpha \in [0, 1)$. We define a homeomorphism $T : X \to X$ by $T(x) = x + \alpha$ (mod 1), i.e.

$$Tx = \begin{cases} x + \alpha & \text{if } 0 \leq x + \alpha \leq 1, \\ x + \alpha - 1 & \text{if } x + \alpha > 1. \end{cases}$$

(An equivalent formulation would be if we let $X = K$ and then define $T : X \to X$ by $T(e^{2\pi i \theta}) = e^{2\pi i(\theta + \alpha)}$. This is equivalent in the sense that the homeomorphism $\rho : \mathbb{R}/\mathbb{Z} \to K$ given by $\rho(t) = e^{2\pi i t}$ relates the two transformations.)

EXAMPLE 3 (SHIFT MAP). For $k \geq 2$ let $X_k = \prod_{n \in \mathbb{Z}} \{1, 2, \ldots, k\}$ denote the space of all sequences taking values $\{1, 2, \ldots, k\}$ indexed by $\mathbb{Z}$. In order to define a metric we first associate to two sequences $x = (x_n)_{n \in \mathbb{Z}}$ and

$x = (x_n)_{n \in \mathbb{Z}}$ an integer $N(x, y) = \min\{N \geq 1 : x_N \neq y_N \text{ or } x_N \neq y_N\}$. We define a metric on X_k by

$$d(x, y) = \begin{cases} \left(\frac{1}{2}\right)^{N(x,y)} & \text{if } x \neq y, \\ 0 & \text{otherwise.} \end{cases}$$

LEMMA 1.1. *X_k is a compact space.*

PROOF. We shall actually show that X_k is sequentially compact. Let $x^{(k)} = (x_n^{(k)})_{n \in \mathbb{Z}}$ $(k = 1, 2, 3, \dots)$ be a sequence in X_k; then we need to show that there exist a point $x \in X_k$ and a sub-sequence $x^{(k_l)} \to x$ $(l = 1, 2, 3, \dots)$.

First observe that the zeroth terms $x_0^{(k)}$ $(k = 1, 2, 3 \dots)$ must take some value in $\{1, 2, \dots, k\}$ infinitely often. Choose such an $x_0 \in \{1, 2, \dots, k\}$ with $x_0^{(k)} = x_0$, for infinitely many m. We continue inductively: For $l > 0$, choose $x_l \in \{1, 2, \dots, k\}$ and $x_{-l} \in \{1, 2, \dots, k\}$ such that $x_{-l}^{(m)} = x_{-l}, \dots, x_0^{(m)} = x_0, \dots, x_l^{(m)} = x_l$, say, for infinitely many m. Finally, we define $x = (x_l)_{l \in \mathbb{Z}}$. For each $l \geq 0$ we choose $m_l := m$ such that $x_{-l}^{(m)} = x_{-l}, \dots, x_0^{(m)} = x_0, \dots, x_l^{(m)} = x_l$; then $d(x^{(m_l)}, x) \leq \frac{1}{2^l}$ and so $d(x^{(m_l)}, x) \to 0$ as $l \to +\infty$. ∎

DEFINITION. We can define a map $\sigma : X_k \to X_k$ by $(\sigma x)_n = x_{n+1}$, $\forall n \in \mathbb{Z}$, i.e.

$$\sigma : (\dots, x_{-2}, x_{-1}, x_0, x_1, x_2, \dots) \longmapsto (\dots, x_{-1}, x_0, x_1, x_2, x_3, \dots).$$

Since this map shifts sequences by one place it is called the *shift map*.

LEMMA 1.2. *The map $\sigma : X_k \to X_k$ is a homeomorphism.*

PROOF. To show continuity we observe that if $x \neq y$ and $d(x, y) = \left(\frac{1}{2}\right)^N$ then we know that $x_i = y_i$ for $-N \leq i \leq N$. Thus we have that $(\sigma x)_i = x_{i+1} = y_{i+1} = (\sigma y)_i$ for $i = -(N+1), \dots, N-1$. This means that $d(\sigma x, \sigma y) \leq \left(\frac{1}{2}\right)^{N-1} = \frac{1}{2} d(x, y)$ and we see that σ is continuous.

Clearly $\sigma : X_k \to X_k$ is invertible (since the inverse transformation $\sigma^{-1} : X_k \to X_k$ simply shifts sequences back one place). Finally, the inverse map $\sigma^{-1} : X_k \to X_k$ is continuous by the same sort of argument as above. ∎

1.2 Transitivity

In this section we shall introduce some basic properties of continuous maps $T : X \to X$ on compact metric spaces X.

DEFINITION. We say that a homeomorphism $T : X \to X$ of a compact metric space X is *transitive* if there exists a point $x \in X$ such that its orbit

$\{T^n x : n \in \mathbb{Z}\} = \{\ldots, T^{-2}x, T^{-1}x, x, Tx, T^2x, \ldots,\}$ is dense in X. We call such a point $x \in X$ a *transitive point*.

We say that a continuous map $T : X \to X$ of a compact metric space X is *(forward) transitive* if there exists a point $x \in X$ such that its *orbit* $\{T^n x : n \in \mathbb{Z}^+\} = \{x, Tx, T^2x, \ldots,\}$ is dense in X. We call such a point $x \in X$ a *(forward) transitive point*.

We can check each of the examples in section 1.1 for this property.

EXAMPLE 1. We shall show that this example is forward transitive when $k = 2$, other cases ebing similar. Consider the sequence $1, 2, 11, 12, 21, 22, 111,$ $112, 121, 122, 221, \ldots, 222, 1111, \ldots$. We can write down $x_n \in \{1, 2\}$, $n \geq 0$, as the nth term in the sequence

$$121112212211111121211222221 \ldots 2221111 \ldots.$$

Finally, consider the point $x \in [0, 1]$ given by the series $x = \sum_{n=0}^{+\infty} \frac{(x_n - 1)}{2^{n+1}}$. We claim that the point x is a (forward) transitive point. Observe that

$$Tx = 2 \left(\sum_{n=0}^{+\infty} \frac{(x_n - 1)}{2^{n+1}} \right) \pmod{1} = x_0 + \sum_{n=0}^{+\infty} \frac{(x_{n+1} - 1)}{2^{n+1}} \pmod{1}$$

$$= \sum_{n=0}^{+\infty} \frac{(x_{n+1} - 1)}{2^{n+1}}.$$

Similarly, $T^k x = \sum_{n=0}^{+\infty} \frac{x_{n+k}}{2^n}$.

To show that the set $\{T^n x : n \geq 0\}$ is dense it suffices to show that for each interval of the form $\left[\frac{p}{2^l}, \frac{p+1}{2^l} \right]$, with $0 \leq p \leq 2^l - 1$, we can find $N \geq 0$ with $T^N x \in \left[\frac{p}{2^l}, \frac{p+1}{2^l} \right]$. Given p we can write it in binary form as $i_0 \ldots i_{n-1}$, with $i_0, \ldots, i_{n-1} \in \{0, 1\}$. But for some N we can find $x_N = i_0, x_{N+1} = i_1, \ldots, x_{N+n-1} = i_{n-1}$. This means that $T^N x \in \left[\frac{p}{2^l}, \frac{p+1}{2^l} \right]$, as required.

EXAMPLE 2. There are two different cases, depending on whether or not α is irrational.

First assume that α is irrational, then the map $T : X \to X$ can be shown to be transitive (and even forward transitive) where $x = 0$, say. It suffices to show that the orbit $\{T^n 0\}_{n \in \mathbb{Z}^+}$ is dense. Since this is an infinite set in $\mathbb{R}/\mathbb{Z}$ we can choose $x \in \mathbb{R}/\mathbb{Z}$ and a sub-sequence $n_i \to +\infty$ with $T^{n_i} 0 = n_i \alpha \pmod{1} \to x$. For any sufficiently small $\epsilon > 0$ we can choose $n_i > n_j$ with $|T^{n_i} 0 - x| < \frac{\epsilon}{2}$ and $|T^{n_j} 0 - x| < \frac{\epsilon}{2}$ and thus $|T^{n_i} 0 - T^{n_j} 0| = |T^{n_i - n_j} 0| < \epsilon$. Moreover, $T^{n_i} 0 \neq T^{n_j} 0$, since if not this would contradict α being irrational. Thus the points $T^{(n_i - n_j)k} 0$, $k \geq 1$, form an ϵ-dense subset of $\mathbb{R}/\mathbb{Z}$. Since ϵ can be chosen arbitarily small this completes the proof of transitivity.

Now assume that $\alpha = \frac{p}{q}$ with $p, q \in \mathbb{Z}$ having no common divisors and $q \neq 0$. For any $x \in X$ the orbit $\{T^n x : n \in \mathbb{Z}\}$ would be a finite set $\{x, x + \frac{1}{q}, \ldots, x + \frac{q-1}{q} \pmod{1}\}$. In particular, T is *not* transitive.

EXAMPLE 3. We shall show that this example is forward transitive. The sequence

$$\{x_n\}_{n\in\mathbb{N}} = \{1, 2, \ldots, k, 1, 1, \ldots, 1, k, 2, 1, \ldots, 2, k, \ldots \underbrace{z_0, z_1, \ldots, z_{N-1}}_{\text{All strings appear}} \ldots\}$$

in X_k (in which all finite strings appear once) is a forward transitive point. To see this choose any point $z \in X_k$ and for any $\epsilon > 0$ choose $N > 0$ sufficiently large that $\left(\frac{1}{2}\right)^N < \epsilon$. If we choose r such that $x_r = z_0, \ldots, x_{r+N-1} = z_{N-1}$ then we see that $(\sigma^r x)_0 = x_r = z_0, \ldots, (\sigma^r x)_{N-1} = x_{r+N-1} = z_{r+N-1}$ and so $d(\sigma^r x, z) \leq \left(\frac{1}{2}\right)^N < \epsilon$.

1.3 Other characterizations of transitivity

The following result gives equivalent conditions for a homeomorphism of a compact metric space to be transitive.

THEOREM 1.3. *The following are equivalent.*

(i) $T : X \to X$ *is transitive.*

(ii) *If U is an open set with $TU = U$ then either U is dense or $U = \emptyset$.*

(iii) *If U, V are non-empty open sets then for some $n \in \mathbb{Z}$ we have that $T^n U \cap V \neq \emptyset$.*

(iv) *The set $\{x \in X :$ the orbit $\{T^n x\}_{n\in\mathbb{Z}}$ is dense in $X\}$ is a dense G_δ set (i.e. the intersection of a countable collection of open dense sets).*

PROOF. (i) $\implies$ (ii). Assume $x \in X$ has a dense orbit. Assume that $TU = U \neq \emptyset$. We can choose $n \in \mathbb{Z}$ such that $T^n x \in U$. Moreover, for any $m \in \mathbb{Z}$ we have that $T^m x \in T^{m-n} U = U$. Since the orbit of x is dense (i.e. $\cup_{m\in\mathbb{Z}} T^m x \subset X$ is dense) we see that U is dense.

(ii) $\implies$ (iii). The T-invariant union $\cup_{n\in\mathbb{Z}} T^n U$ is dense in X by assumption (ii). Thus $\cup_{n\in\mathbb{Z}} T^n U \cap V \neq \emptyset$ and so $\exists n \in \mathbb{Z}$ with $T^n U \cap V \neq \emptyset$.

(iii) $\implies$ (iv). Consider a dense set $\{x_n\}_{n\in\mathbb{N}}$ and consider the balls of radius $\frac{1}{k}, k \geq 1$, denoted by $B\left(x_n, \frac{1}{k}\right)$. We can identify

$$\{x \in X : \{T^m x\}_{m\in\mathbb{Z}} \text{ is dense in } X\} = \cap_{n=0}^{+\infty} \cap_{k=1}^{+\infty} \cup_{m=-\infty}^{+\infty} T^m B\left(x_n, \frac{1}{k}\right)$$

(i.e. $\forall n \geq 0, \forall k \geq 1, \exists m \in \mathbb{Z}$ with $T^m x \in B\left(x_n, \frac{1}{k}\right)$).

(iv) $\implies$ (i). This is immediate.   ∎

REMARK. There is a similar result giving equivalent conditions for forward transitivity [4, p. 128]

1.4 Transitivity for subshifts of finite type

In section 1.1 we defined the shift transformation $\sigma : X_k \to X_k$ on $X_k = \prod_{n \in \mathbb{Z}} \{1, \ldots, k\}$. For any closed σ-invariant subset $X \subset X_k$ (i.e. $\sigma(X) = X$) we consider the restriction $\sigma|_X$. We can use the same notation $\sigma : X \to X$.

DEFINITION. Let A be a $k \times k$ matrix with entries 0 or 1. We call the matrix *irreducible* if $\forall 1 \leq i, j, \leq k$, $\exists N > 0$ such that $A^N(i,j) > 0$.

EXAMPLES. When $k = 3$ the matrix $A = \begin{pmatrix} 0 & 1 & 1 \\ 0 & 1 & 1 \\ 1 & 0 & 0 \end{pmatrix}$ is irreducible. However, the matrix $A' = \begin{pmatrix} 1 & 1 & 0 \\ 1 & 1 & 0 \\ 0 & 0 & 1 \end{pmatrix}$ is not irreducible. (These properties are readily checked).

DEFINITION. Given a $k \times k$ matrix A with entries 0 or 1 we define

$$X_A = \{(x_n)_{n \in \mathbb{Z}} \in \prod_{n=-\infty}^{\infty} \{1, \ldots, k\} : A(x_n, x_{n+1}) = 1, n \in \mathbb{Z}\}.$$

We define the *subshift of finite type* $\sigma : X_A \to X_A$ to be the restriction $\sigma|X_A$.

The following gives necessary and sufficient conditions for $\sigma : X_A \to X_A$ to be transitive.

THEOREM 1.4. *A subshift of finite type $\sigma : X_A \to X_A$ is transitive if and only if A is irreducible.*

PROOF. Assume that σ is transitive. Consider the sets

$$[i]_0 := \{(x_n)_{n \in \mathbb{Z}} \in X_A : x_0 = i\}$$

for $i = 1, \ldots, k$. These sets are open. Given $1 \leq i, j \leq k$ we know that there exists $N > 0$ such that $\sigma^{-N}[j]_0 \cap [i]_0 \neq \emptyset$. Choose $(x_n)_{n \in \mathbb{Z}} \in \sigma^{-N}[j]_0 \cap [i]_0$; then we know that $x_0 = i$ and $x_N = j$. Notice that

$$A^N(i,j) = \sum_{r_1=1}^{k} \cdots \sum_{r_{N-1}=1}^{k} A(i, r_1) A(r_1, r_2) \ldots A(r_{N-2}, r_{N-1}) A(r_{N-1}, j).$$

But since $A(i, x_1) = A(x_1, x_2) = \ldots = A(x_{N-1}, j) = 1$ we see that $A^N(i,j) \geq 1$.

Conversely, assume that for $1 \leq i, j \leq k$ we have that $A^N(i,j) \geq 1$. Given $U, V \neq \emptyset$ open sets we can choose $(i_n)_{n \in \mathbb{Z}} \in U$ and $(j_n)_{n \in \mathbb{Z}} \in V$ such that for $M > 0$ sufficiently large

$$U \supset [i_{-M}, i_{-M-1}, \ldots, i_M]_{-M}^{M} := \{(x_n)_{n \in \mathbb{Z}} \in X_A : x_k = i_k, -M \leq k \leq M\},$$

$$V \supset [j_{-M}, j_{-M-1}, \ldots, j_M]_{-M}^{M} := \{(x_n)_{n \in \mathbb{Z}} \in X_A : x_k = j_k, -M \leq k \leq M\}.$$

By hypothesis we can find $N > 0$ such that $A^N(i_M, j_{-M}) \geq 1$. This means that we can find a string $x'_1, \ldots, x'_{N-1}$ such that $A(i_M, x'_1) = A(x'_1, x'_2) = \ldots = A(x'_{N-1}, j_{-M}) = 1$ and then define

$$x_n = \begin{cases} i_n & \text{if } n \leq M, \\ x'_{n-M} & \text{if } M+1 \leq n \leq M+N-1, \\ j_{n-(2M+N)} & \text{if } M+N \leq n; \end{cases}$$

then we have that $x \in U \cap \sigma^N V$ i.e. $U \cap \sigma^N V \neq \emptyset$.   ∎

1.5 Minimality and the Birkhoff recurrence theorem

In this section we want to present a simple but important recurrence result, called the *Birkhoff recurrence theorem*. Our starting point is to define the following property.

DEFINITION. A homeomorphism $T : X \to X$ is *minimal* if for every $x \in X$ the orbit $\{T^n x : n \in \mathbb{Z}\}$ is dense in X.

The following is obvious from the definitions

PROPOSITION 1.5. *A minimal homeomorphism is necessarily transitive.*

We can now consider each of the examples from section 1.1 and ask which of these are minimal. Since Example 1 is not a homeomorphism we begin with Example 2.

EXAMPLE 2.

LEMMA 1.6. *When α is irrational then $T(x) = x + \alpha$ is minimal.*

PROOF. It suffices to show that for every $x \in \mathbb{R}/\mathbb{Z}$ and every neighbourhood $(y - \epsilon, y + \epsilon)$ $(y \in \mathbb{R}/\mathbb{Z}, \epsilon > 0)$ we can find $n \geq 1$ such that $T^n(x) \in (y - \epsilon, y + \epsilon)$.

We already know that T is transitive (i.e. there exists at least one transitive point $x_0 \in \mathbb{R}/\mathbb{Z}$ with dense orbit). Fix $y \in \mathbb{R}/\mathbb{Z}$ and use the transitivity to choose a sub-sequence n_i with $T^{n_i} x_0 \to (y - x + x_0)$ as $i \to +\infty$. Thus

$$T^{n_i} x = n_i \alpha + x \quad (\text{mod } 1)$$
$$= n_i \alpha + x_0 + (x - x_0) \quad (\text{mod } 1)$$
$$= T^{n_i}(x_0) + (x - x_0) \quad (\text{mod } 1)$$
$$\to y + (x_0 - x) + (x - x_0) = y \quad (\text{mod } 1).$$

                    ∎

EXAMPLE 3. The shift map is not minimal since it contains a fixed point (e.g. $x = (\ldots, 1, 1, 1, \ldots)$).

The following theorem gives equivalent definitions.

THEOREM 1.7. *Let $T : X \to X$ be a homeomorphism of a compact metric space. The following properties are equivalent.*

(i) *T is minimal.*

(ii) *If $TE = E$ is a closed T-invariant set, then either $E = \emptyset$ or $E = X$.*

(iii) *If $U \neq \emptyset$ is an open set then $X = \cup_{n \in \mathbb{Z}} T^n U$.*

PROOF. *(i)* $\implies$ *(ii)* Assume that $TE = E \neq \emptyset$ and choose $x \in E$. Hypothesis (i) gives that $X = \mathrm{cl}\,(\{T^n x\}_{n \in \mathbb{Z}}) \subset E \subset X$.

(ii) $\implies$ *(iii)* Given a non-empy open set U let $E = X - (\cup_{n \in \mathbb{Z}} T^n U)$. By construction $TE = E$ and $E \neq X$ (since $U \neq \emptyset$) and so by hypothesis (ii) we have that $E = \emptyset$. Thus $X = \cup_{n \in \mathbb{Z}} T^n U$.

(iii) $\implies$ *(i)* Fix $x \in X$ and an open neighbourhood $U \subset X$. Since $x \in T^n U$ for some $n \in \mathbb{Z}$ (by hypothesis (iii)) we have that $T^{-n} x \in U$. This shows that the orbit $\{T^n x\}_{n \in \mathbb{Z}}$ is dense in X.
∎

Using property (ii) we get the following suprising result that every homeomorphism contains a minimum homeomorphism.

THEOREM 1.8. *Let $T : X \to X$ be a homeomorphism of a compact metric space X. There exists a non-empty closed set $Y \subset X$ with $TY = Y$ and $T : Y \to Y$ is minimal.*

PROOF. This follows from an application of Zorn's Lemma. Let $\mathcal{E} = \{Z \subset X : TZ = Z\}$ denote the family of all T-invariant subsets of X with the partial ordering by inclusion, i.e. $Z_1 \leq Z_2$ iff $Z_1 \subset Z_2$.

Every totally ordered subset (or "chain") $\{Z_\alpha\}$ has a least element $Z = \cap_\alpha Z_\alpha$ (which is non-empty by compactness of X). Thus by Zorn's lemma there exists a minimal element $Y \subset X$ (i.e. $Y \in \mathcal{E}$ and $Y' \in \mathcal{E}$ with $Y' \leq Y$ implies that $Y = Y'$). By property (ii) of Theorem 1.7 this can be re-interpreted as saying thet $T : Y \to Y$ is minimal.
∎

As a corollary we get the following simple but elegant result.

COROLLARY 1.8.1 (BIRKHOFF RECURRENCE THEOREM). *Let $T : X \to X$ be a homeomorpism of a compact metric space X. We can find $x \in X$ such that $T^{n_i} x \to x$ for a sub-sequence of the integers $n_i \to +\infty$.*

PROOF. By Theorem 1.8 we can choose a T-invariant subset $Y \subset X$ such that $T : Y \to Y$ is minimal. For any $x \in Y \subset X$ we have the required property.
∎

EXAMPLE 2. Consider the case $X = \mathbb{R}/\mathbb{Z}$ and $T : X \to X$ defined by $Tx = x + \alpha \pmod 1$, where α is an irrational number.

Let $\epsilon > 0$; then we can find $n > 0$ (by Birkhoff's theorem) such that $|\alpha n \ (\text{mod } 1)| \leq \epsilon$, i.e. there exists $p \in \mathbb{N}$ such that $-\epsilon \leq \alpha n - p \leq \epsilon$. Rewriting this, we have that for any irrational α, $\exists p, n \in \mathbb{N}$ such that $|\alpha - \frac{p}{n}| \leq \frac{\epsilon}{n}$. This is a (marginal) improvement on the most obvious estimate.

1.6 Commuting homeomorphisms

Let $T_1, \ldots, T_N : X \to X$ be commuting homeomorphisms on a compact metric space X, i.e. $T_i T_j = T_j T_i$ for $1 \leq i, j \leq N$. In this section we shall briefly consider how some of the ideas from the previous section might be modified for such families of maps.

EXAMPLE 4. Consider the simple example of two rotations on the torus $X = \mathbb{R}^n / \mathbb{Z}^n$ of the form

$$T_1(x_1, \ldots, x_n) = (x_1 + a_1^{(1)}, \ldots, x_n + a_n^{(1)}) \ (\text{mod } 1),$$

$$\vdots$$

$$T_N(x_1, \ldots, x_n) = (x_1 + a_1^{(N)}, \ldots, x_n + a_n^{(N)}) \ (\text{mod } 1),$$

where $(a_1^{(1)}, \ldots, a_n^{(1)}), \ldots, (a_1^{(N)}, \ldots, a_n^{(N)}) \in \mathbb{R}^n$.

We can consider all closed *simultaneously invariant* sets $A \subset X$, i.e. $T_i A = A$, $i = 1, \ldots, N$. By a similar argument to that before, we can consider the partial order by inclusion on all such closed sets and by applying Zorn's lemma (just as in the proof of Theorem 1.8) we can deduce that there exists a closed set $X_0 \subset X$ such that

   (i) $T_i X_0 = X_0$, $i = 1, \ldots, N$.
   (ii) whenever $A \subset X_0$ with A closed and $T_i A = A$ for $i = 1, \ldots, N$ then necessarily $A = X_0$.

The following lemma will prove useful in chapter 2.

LEMMA 1.9. *For each open set $U \subset X_0$ we can choose a finite M and $n_{ij} \in \mathbb{Z}$ with $1 \leq i \leq N, 1 \leq j \leq M$ with $X_0 = \cup_{j=1}^{M} (T_1^{n_{1j}} \circ \ldots \circ T_N^{n_{Nj}}) U$.*

PROOF. Clearly $X_0 = \cup_{n_1 \in \mathbb{Z}} \ldots \cup_{n_N \in \mathbb{Z}} (T_1^{n_1} \circ \ldots \circ T_N^{n_N}) U$ (since otherwise the difference $X_0 - (\cup_{n_1 \in \mathbb{Z}} \ldots \cup_{n_N \in \mathbb{Z}} (T_1^{n_1} \circ \ldots \circ T_N^{n_N}) U)$ is a closed (non-empty) set invariant under $T_1, \ldots, T_N$, contradicting property (ii) above). Now by compactness we can choose a *finite* subcover. This completes the proof. ∎

To formulate a generalisation of the Birkhoff recurrence theorem to a family of commuting maps is a more substantial exercise, and will be a principal part of chapter 2.

1.7 Comments and references

A wealth of interesting examples can be found in the literature (cf. [1], [2], [3], [4], [5]).

The simple Birkhoff recurrence theorem has a version for commuting homeomorphisms (the multiple Birkhoff recurrence theorem) which we shall describe in Chapter 2. The corresponding result to the Birkhoff recurrence theorem in ergodic theory is the Poincaré recurrence theorem, which appears in section 9.2.

References

1. P. Billingsley, *Ergodic Theory and Information*, Addison-Wesley, New York, 1965.
2. R. Devaney, *An Introduction to Chaotic Dynamical Systems*, Addison-Wesley, New York, 1989.
3. A. Katok and B. Hasselblatt, *Introduction to the Modern Theory of Dynamical Systems*, C.U.P., Cambridge, 1995.
4. W. Szlenk, *An Introduction to the Theory of Smooth Dynamical Systems*, Wiley, New York, 1984.
5. P. Walters, *An Introduction to Ergodic Theory*, Graduate Texts in Mathematics, 79, Springer, Berlin, 1982.

CHAPTER 2

AN APPLICATION OF RECURRENCE
TO ARITHMETIC PROGRESSIONS

In this chapter we shall describe a particularly nice application of the recurrence ideas from chapter 1 to a result in number theory.

2.1 Van der Waerden's theorem

We begin with a simple idea from number theory.

DEFINITION. An *arithmetic progression* is a sequence of integers $\{a + jb\}_{j=0}^{N-1}$ for $a, b \in \mathbb{Z}$ ($b \neq 0$), $N \geq 1$. We call N the *length* of the arithmetic progression.

EXAMPLES.

(1) The sequence $10, 13, 16, 19, 22$ is an arithmetic progression with $a = 10, b = 3, N = 5$.
(2) The sequence $-4, 0, 4, 8$ is an arithmetic progression with $a = -4, b = 4, N = 4$.

Consider a *partition* of the integers $\mathbb{Z} = B_1 \cup \ldots \cup B_l$ where

(i) $B_i \neq \emptyset$,
(ii) $B_i \cap B_j = \emptyset$ for $i \neq j$.

The main result we want to prove is the following.

THEOREM 2.1 (VAN DER WAERDEN). *Consider a finite partition* $\mathbb{Z} = B_1 \cup \ldots \cup B_k$. *At least one element* B_r *in the partition will contain arithmetic progressions of arbitrary length (i.e.* $\exists 1 \leq r \leq k$, $\forall N > 0$, $\exists a, b \in \mathbb{Z}$ ($b \neq 0$) *such that* $a + jb \in B_r$ *for* $j = 0, \ldots, N - 1$).

Since an arithmetic progression of length N contains arithmetic progressions of all shorter lengths, this is equivalent to: $\exists N_i \to +\infty$, $\exists a_i, b_i \in \mathbb{Z}$ such that $a_i + jb_i \in B_r$ for $j = 0, \ldots, N_i - 1$.

We give below some simple examples.

EXAMPLES.

(1) If the sets $B_2, \ldots, B_k$, say, in the partition are finite then it is easy to see that B_1 is the element with arithmetic progressions of arbitrary length.

(2) If $\mathbb{Z} = B_1 \cup B_2$ where $B_1 = \{$odd numbers$\}$ and $B_2 = \{$even numbers$\}$ then both contain arithmetic progressions of arbitrary length.

(3) If $B_1 = \{$prime numbers$\}$ and $B_2 = \{$non-prime numbers$\}$ then B_2 contains arithmetic progressions of arbitrary length. However, it is an unsolved problem as to whether B_1 contains arithmetic progressions of arbitrary length.

HISTORICAL NOTE. This result was originally conjectured by Baudet and proved by Van der Waerden in 1927 [6, 7]. The theorem gained a wider audience when it was included in Khintchine's famous book *Three pearls in number theory* [4]. The dynamical proof we give is due to Furstenberg and Weiss [3](from 1978).

2.2 A dynamical proof

The key to proving Van der Waerden's theorem is the following generalization of Birkhoff's theorem.

THEOREM 2.2. *Let $T_1, \ldots, T_N : X \to X$ be homeomorphisms of a compact metric space such that $T_i T_j = T_j T_i$ for $1 \leq i, j \leq N$. There exist $x \in X$ and $n_j \to +\infty$ such that $d(T_i^{n_j} x, x) \to 0$ for each $i = 1, \ldots, N$.*

We shall first prove Theorem 2.1 assuming Theorem 2.2 and then return to the proof of Theorem 2.2.

PROOF OF THEOREM 2.1 (ASSUMING THEOREM 2.2). We want to begin by associating to the partition $\mathbb{Z} = B_1 \cup \ldots \cup B_k$ a suitable homeomorphism $T : X \to X$ (and then we set $T_j = T^j$, $j = 1, \ldots, N$).

Let $\Omega = \prod_{n \in \mathbb{Z}} \{1, \ldots, k\}$ and then we can associate to the partition $\mathbb{Z} = B_1 \cup \ldots \cup B_k$ a sequence $z = (z_n)_{n \in \mathbb{Z}} \in \Omega$ by $z_n = i$ if and only if $n \in B_i$.

Let $\sigma : \Omega \to \Omega$ be the shift introduced in Example 3 of section 1.1 (i.e. $(\sigma x)_n = x_{n+1}$, $n \in \mathbb{Z}$). Consider the orbit $\{\sigma^n z : n \in \mathbb{Z}\}$ and its closure $X = \mathrm{cl}\,(\cup_{n \in \mathbb{Z}} \sigma^n z)$. Finally, we define $T_i := T^i = (T \circ \ldots \circ T)$ (T composed with itself i times).

By Theorem 2.2 (with $\epsilon = \frac{1}{4}$) we can find $x \in X$ and $b \geq 1$ with

$$d(T_1^b x, x) < \frac{1}{4}, d(T_2^b x, x) < \frac{1}{4}, \ldots, d(T_N^b x, x) < \frac{1}{4}.$$

Since $X = \mathrm{cl}\,(\cup_{n \in \mathbb{Z}} \sigma^n z)$ we can choose $a \in \mathbb{Z}$ such that

$$d(x, T^a z) < \frac{1}{4}, d(T_1^b x, T^a T_1^b z) < \frac{1}{4}, \ldots, d(T_N^b x, T^a T_N^b z) < \frac{1}{4}.$$

Thus, for each $i = 1, \ldots, N$ we have that

$$d(T^a T_i^b x, T^a z) \leq d(T^a T_i^b x, T_i^b x) + d(T_i^b x, x) + d(x, T^a z) < \frac{1}{4} + \frac{1}{4} + \frac{1}{4} = \frac{3}{4}.$$

Since $d(x,y) = \left(\frac{1}{2}\right)^{N(x,y)}$ (where $N(x,y) = \min\{|N| \geq 1 : x_N \neq y_N$, or $x_{-N} \neq y_{-N}\}$) we see that $(T^a T_i^b x)_0 = x_{b+ia} = z_a \in \{1,\ldots,k\}$ for $i = 1,\ldots,N$. This means that $b+ia \in B_{z_0}$, for $i = 1,\ldots,N$, and completes the proof of Theorem 2.1.

∎

All that remains is to prove Theorem 2.2. This is a fairly detailed proof and we to help clarify matters we shall divide it into sublemmas.

PROOF OF THEOREM 2.2. We shall use a proof by induction.

CASE $N = 1$. For $N = 1$ the multiple Birkhoff recurrence theorem reduces to the (usual) Birkhoff recurrence theorem (Corollary 1.8.1).

INDUCTIVE STEP. Assume that the result is known for $N-1$ commuting homeomorphisms. We need to show that it holds for N commuting homeomorphisms.

SIMPLIFYING FACT. We can assume that X is the *smallest* closed set invariant under each of $T_1,\ldots,T_N$. If this is not the case we can restrict to such a set (using Zorn's lemma as in section 1.6).

In order to establish the Birkhoff multiple recurrence theorem for these N commuting homeomorphisms, the following simple alternative formulation of this result is useful.

ALTERNATIVE FORMULATION. Let $\mathcal{X}_N = X \times \ldots \times X$ be the N-fold cartesian product of X and let $\mathcal{D}_N = \{(x,\ldots,x) \in \mathcal{X}_N\}$ be the diagonal of the space. Let $S : \mathcal{X}_N \to \mathcal{X}_N$ be given by $S(x_1,\ldots,x_N) = (T_1 x_1,\ldots,T_N x_N)$. Then the following are equivalent:

$(i)_N$ the Birkhoff multiple recurrence holds for $T_1,\ldots,T_N$;

$(ii)_N$ $\exists \underline{z} = (z,\ldots,z) \in \mathcal{D}_N$ such that $d_{\mathcal{X}_N}(S^{n_i}\underline{z},\underline{z}) \to 0$ as $n_i \to +\infty$ (where $d_{\mathcal{X}_N}(\underline{z},\underline{w}) = \sup_{1 \leq i \leq N} d(z_i,x_i)$).

We can apply the inductive hypothesis to the $(N-1)$ commuting homeomorphisms $T_1 T_N^{-1},\ldots,T_{N-1}T_N^{-1}$ and using the equivalence of $(i)_{N-1}$ and $(ii)_{N-1}$ above we have that for the map $R := T_1 T_N^{-1} \times \ldots \times T_{N-1} T_N^{-1} : \mathcal{X}_{N-1} \to \mathcal{X}_{N-1}$ defined by

$$R(x_1,\ldots,x_{N-1}) \mapsto (T_1 T_N^{-1} x_1,\ldots,T_{N-1}T_N^{-1}x_{N-1})$$

there exists $\underline{z} = (z,\ldots,z) \in \mathcal{D}_{N-1} \subset \mathcal{X}_{N-1}$ with $d_{\mathcal{X}_{N-1}}(R^{n_i}\underline{z},\underline{z}) \to 0$ as $n_i \to +\infty$. In particular, $d_{\mathcal{X}_N}(S^{n_i}\mathcal{D}_N,\mathcal{D}_N) \to 0$ as $n_i \to +\infty$ (since both $\underline{z} = (z,\ldots,z), \underline{z}' = (T_N^{-n_i}z,\ldots,T_N^{-n_i}z) \in \mathcal{D}_N$).

Thus we have proved the following result.

SUBLEMMA 2.2.1. $\forall \epsilon > 0$, $\exists z, z' \in \mathcal{D}_N$, $\exists n \geq 1$ such that $d_{\mathcal{X}_N}(S^n z, z') < \epsilon$.

Unfortunately, this is not quite in the form of $(ii)_N$ we need for the inductive step. (For example, we would like to take $z = z'$.) To get a stronger result, we break the argument up into steps represented by the following sublemmas.

SUBLEMMA 2.2.2. $\forall \epsilon > 0, \forall x \in \mathcal{D}_N, \exists y \in \mathcal{D}_N$ and $\exists n \geq 1$ such that $d(S^n y, x) < \epsilon$.

(This changes one of the quantifiers $\exists$ to $\forall$.)

SUBLEMMA 2.2.3. $\forall \epsilon > 0$, $\exists z \in \mathcal{D}_N$ and $n \geq 1$ such that $d(S^n z, z) < \epsilon$

(This is almost the Birkhoff multiple recurrence theorem, except that z might still depend on the choice of $\epsilon > 0$.)

We will now complete the proof of the Birkhoff multiple recurrence theorem assuming Sublemma 2.2.3. (We shall then return to the proofs "Sublemma 2.2.1 $\implies$ Sublemma 2.2.2" and "Sublemma 2.2.2 $\implies$ Sublemma 2.2.3" in the next section.)

Consider the function $F : \mathcal{D}_N \to \mathbb{R}^+ = [0, +\infty)$ defined by $F(x) = \inf_{n \geq 1} d(S^n x, x)$. It is easy to see that to complete the proof of Theorem 2.2 we need only show there exists a point $x_0 \in \mathcal{D}_N$ with $F(x_0) = 0$. To show this fact, the following properties of F are needed.

SUBLEMMA 2.2.4.

(i) $F : \mathcal{D} \to \mathbb{R}^+$ is upper semi-continuous (i.e. $\forall x \in \mathcal{D}_N, \forall \epsilon > 0, \exists \delta > 0$ such that $d(x, y) < \delta \implies F(y) \leq F(x) + \epsilon$).

(ii) $\exists x_0 \in \mathcal{D}_N$ such that $F : \mathcal{D}_N \to \mathbb{R}^+$ is continuous at x_0.

PROOF.

(i) This is an easy exercise from the definition of F.

(ii) For $\epsilon > 0$ we can define $A_\epsilon = \{x \in \mathcal{D} : \forall \eta > 0, \exists y \text{ such that } d(y, x) < \eta \text{ and } F(y) \leq F(x) - \epsilon\}$ (i.e. $\exists$ points y arbitrarily close to x with $F(y) \leq F(x) - \epsilon$). Notice that

(a) A_ϵ is closed,

(b) A_ϵ has empty interior.

(To see part (b) observe that if $\text{int}(A_\epsilon) \neq \emptyset$ we could choose a sequence of pairs $x, x_1 \in \text{int}(A_\epsilon)$ with $F(x_1) \leq F(x) - \epsilon$, $x_1, x_2 \in \text{int}(A_\epsilon)$ with $F(x_2) \leq F(x_1) - \epsilon$, etc. Together these inequalities give $F(x_n) \leq F(x) - n\epsilon < 0$ for n arbitrarily large. But this contradicts $F \geq 0$).

The set of points at which F is continuous is

$$\{y \in \mathcal{D} : x \notin A_\epsilon, \epsilon > 0\} = \cap_{n=1}^{\infty} \left(\mathcal{D} - A_{\frac{1}{n}} \right).$$

Since this is a countable intersection of open dense sets, it is still dense (by Baire's theorem). Thus there exists at least one point of continuity for $F : \mathcal{D} \to \mathbb{R}^+$ (in fact, infinitely many). This completes the proof of Sublemma 2.2.4.

∎

Let x_0 be such a point of continuity.

Assume for a contradiction that $F(x_0) > 0$. We can then choose $\delta > 0$ and an open neighbourhood $U \ni x_0$ such that $F(x) > \delta > 0$ for $x \in U$. However, we also know that

$$\mathcal{D}_N \subset \cup_{j=1}^M \left(T_1^{n_{1j}} \circ \ldots \circ T_N^{n_{Nj}} \right) U$$

(since by the simplifying assumption X is the smallest closed set invariant under $T_1, \ldots, T_N$ and so we may apply Lemma 1.9 from Chapter 1).

By (uniform) continuity of the family $\{T_1^{n_{1j}} \circ \ldots \circ T_N^{n_{Nj}}\}_{j=1}^M$ there exists $\eta > 0$ such that

$$d(x,y) < \eta \implies d(T_1^{n_{1j}} \circ \ldots \circ T^{n_{Nj}}x, T_1^{n_{1j}} \circ \ldots \circ T_N^{n_{Nj}}y) < \delta \qquad (2.1)$$

(for $1 \le j \le M$) Observe that for $y \in \left(T_1^{n_{1j}} \circ \ldots \circ T_N^{n_{Nj}} \right)^{-1} U$ $(j = 1, \ldots, M)$ we have that $F(y) \ge \eta$. If this were not the case then there would exist $n \ge 1$ with $d(y, T^n y) < \eta$, from the definition of F. This then implies that $d(T_1^{n_{1j}} \circ \ldots \circ T_N^{n_{Nj}}y, T_1^{n_{1j}} \circ \ldots \circ T_N^{n_{Nj}}T^n y) < \delta$ by (2.1). Choosing $x := T_1^{n_{1j}} \circ \ldots \circ T_N^{n_{Nj}}y \in U$ gives $F(x) = \inf_{n \ge 1} d(x, T^n x) < \delta$ which contradicts our hypothesis.

Finally we see that by (2.1) we have $F(y) \ge \eta$ for all $y \in \mathcal{D}_N$. However, this contradicts Sublemma 2.2.3 and we conclude that $F(x_0) = 0$

The proof of Theorem 2 is finished (given the proofs of Sublemma 2.2.2 and Sublemma 2.2.3).

∎

2.3. The proofs of Sublemma 2.2.2 and Sublemma 2.2.3

We now supply the missing proofs of Sublemma 2.2.2 and Sublemma 2.2.3.

PROOF OF SUBLEMMA 2.2.2 (ASSUMING SUBLEMMA 2.2.1). Consider the N commuting maps $\hat{T}_1, \hat{T}_2, \ldots, \hat{T}_N : \mathcal{D}_N \to \mathcal{D}_N$ defined by

$$\begin{cases} \hat{T}_1 = T_1 \times \ldots \times T_1 : \mathcal{D}_N \to \mathcal{D}_N, \\ \hat{T}_2 = T_2 \times \ldots \times T_2 : \mathcal{D}_N \to \mathcal{D}_N, \\ \quad \ldots \\ \hat{T}_N = T_N \times \ldots \times T_N : \mathcal{D}_N \to \mathcal{D}_N. \end{cases}$$

We want to apply Lemma 1.9 to these commuting maps with the choice of
open set $U = \{w \in \mathcal{D}_N : d_{\mathcal{D}_N}(x, w) < \frac{\epsilon}{2}\}$. This allows us to conclude that
there exist $n_{1j}, \ldots, n_{Nj}$ $(j = 1, \ldots, M)$ such that

$$\mathcal{D}_N = \cup_{j=1}^N \hat{T}^{-n_{1j}} \ldots \hat{T}^{-n_{Nj}} U.$$

Thus for any $z \in \mathcal{D}_N$ we have some $1 \le j \le M$ such that

$$d_{\mathcal{D}_N}(\hat{T}^{n_{1j}} \ldots \hat{T}^{n_{Nj}} z, x) < \frac{\epsilon}{2}. \tag{2.2}$$

Next we can use (uniform) continuity of $T^{n_{1j}} \circ \ldots \circ \hat{T}^{n_{Nj}}$ to say that there
exsits $\delta > 0$ such that whenever $d(z, z') < \delta$ for $z, z' \in \mathcal{D}_N$ then we have that

$$d_{\mathcal{D}_N}(\hat{T}^{-n_{1j}} \circ \ldots \circ \hat{T}^{-n_{Nj}} z, \hat{T}^{-n_{1j}} \circ \ldots \circ \hat{T}^{-n_{Nj}} z') < \frac{\epsilon}{2}. \tag{2.3}$$

By Sublemma 2.2.1 $\exists z, z' \in \mathcal{D}_N$ and $\exists n \ge 1$ such that $d_{\mathcal{D}_N}(T^n z, z') < \delta$.
Therefore by inequality (2.3) we have that

$$d_{\mathcal{D}_N}\left(T^n\left(\hat{T}^{n_{1j}} \circ \ldots \circ \hat{T}^{n_{Nj}} z\right), \hat{T}^{n_{1j}} \circ \ldots \circ \hat{T}^{n_{Nj}} z'\right) < \frac{\epsilon}{2}. \tag{2.4}$$

Writing $y = \hat{T}^{n_{1j}} \ldots \hat{T}^{n_{Nj}} z$ and comparing (2.2) and (2.4) gives that

$$d_{\mathcal{D}_N}(T^n y, x) \le d_{\mathcal{D}_N}(T^n y, \hat{T}^{n_{1j}} \circ \ldots \circ \hat{T}^{n_{Nj}} z') + d_{\mathcal{D}_N}(\hat{T}^{n_{1j}} \circ \ldots \circ \hat{T}^{n_{Nj}} z', x)$$
$$< \frac{\epsilon}{2} + \frac{\epsilon}{2} = \epsilon.$$

This completes the proof of Sublemma 2.2.2. ∎

PROOF OF SUBLEMMA 2.2.3 (ASSUMING SUBLEMMA 2.2.2). Fix $z_0 \in \mathcal{D}_N$
and let $\epsilon_1 = \frac{\epsilon}{2}$. By Sublemma 2.2.2 we can choose $n_1 \ge 1$ and $z_1 \in \mathcal{D}_N$ with
$d(T^{n_1} z_1, z_0) < \epsilon_1$.

By continuity of T^{n_1} we can find $\epsilon_1 > \epsilon_2 > 0$ such that $d(z, z_1) < \epsilon_2$
implies that $d(T^{n_1} z, z_0) < \epsilon_1$.

We can now continue inductively (for $k \ge 2$):

(a) By Sublemma 2.2 we can choose $n_k \ge 1$ and $z_k \in \mathcal{D}_N$ with $d(T_k^n z_k, z_{k-1}) < \epsilon_k$.
(b) By continuity of T^{n_k} we can find $\epsilon_k > \epsilon_{k+1} > 0$ such that $d(z, z_k) < \epsilon_{k+1}$ implies that $d(T^{n_k} z, z_{k-1}) < \epsilon_k$.

This results in sequences

$$z_0, z_1, z_2, \ldots \in \mathcal{D}_N,$$
$$n_0, n_1, n_2 \ldots \in \mathbb{N}, \quad \text{such that} \left\{ \begin{array}{l} d(T^{n_k} z_k, z_{k-1}) < \epsilon_k, k \leq 1, \\ d(z, z_i) < \epsilon_{k+1} \implies d(T^{n_k} z, z_{k-1}) < \epsilon_k. \end{array} \right.$$
$$\epsilon_0 > \epsilon_1 > \epsilon_2 > \ldots$$

In particular we get that whenever $j < i$ then

$$d(T^{n_i + n_{i-1} + \ldots + n_{j+2} + n_{j+1}} z_i, z_j) < \epsilon_{i+1} \leq \frac{\epsilon}{2}. \tag{2.5}$$

By compactness of $\mathcal{D}$ we can find $d(z_i, z_j) < \frac{\epsilon}{2}$ for some $j < i$.
By the triangle inequality we have that for $N = n_i + n_{i-1} + \ldots + n_{j+1}$

$$d(T^N z_i, z_i) \leq d(T^N z_i, z_j) + d(z_j, z_i) < \epsilon.$$

Thus the choice $z = z_i$ completes the proof of Sublemma 2.2.3. ∎

2.4 Comments and references

A treatment of Van der Waerden's theorem (and many other related applications of dynamics to number theory) can be found in [1]. The proof originally appeared in the article [3] and the survey [2]. An account also appears in [4, 4.3B].

Sublemma 2.2.2 was originally proved by Bowen.

Finally, there is a stronger version of this result due to Szemeredi. In chapter 16 we shall present Furstenberg's proof of this using ergodic theory.

References

1. H. Furstenburg, *Recurrence in Ergodic Theory and Combinatorial Number Theory*, P.U.P., Princeton N.J., 1981.
2. H.Furstenburg, *Poincaré recurrence and number theory*, Bull. Amer. Math. Soc. **5** (1981), 211-234.
3. H. Furstenburg and B. Weiss, *Topological dynamics and combinatorial number theory*, J. d'Analyse Math. **34** (1978), 61-85.
4. A. Khintchine, *Three pearls of number theory*, Graylock Press, New York, 1948.
5. K. Petersen, *Ergodic Theory*, C.U.P., Cambridge, 1983.
6. B.L. Van der Waerden, *Beweis einer Baudet'schen Vermutung*, Nieuw. Arch. Wisk. **15** (1927), 212-16.
7. B.L. Van der Waerden, *How the proof of Baudet's conjecture was found*, Studies in Pure Mathematics presented to Richard Rado (L. Mirsky, ed.), Academic Press, London, 1971, pp. 251-260.

CHAPTER 3

TOPOLOGICAL ENTROPY

In this chapter we shall introduce an important numerical quantity called topological entropy. This is an important quantifier of their dynamical behaviour (as we shall see in chapters 4 and 5). It also plays an important rôle as an invariant for the classification of continuous maps up to conjugacy.

3.1 Definitions

We begin with some simple ideas on finite open covers for X.

DEFINITIONS. Let X be a compact metric space. If $\alpha = \{A_i\}, \beta = \{B_j\}$ are (finite) open covers of X, define the *refinement* $\alpha \vee \beta = \{A_i \cap B_j : A_i \cap B_j \neq \emptyset\}$. More generally, if $\alpha^r = \{A_1^r, \dots, A_{N_r}^r\}$, $r = 1, \dots, k$, are open covers of X (of cardinality N_r) then we define their *refinement*

$$\vee_{r=1}^k \alpha^r = \{A_{i_1}^1 \cap A_{i_2}^2 \cap \dots \cap A_{i_k}^k : i_j \in \{1, \dots, N_r\}, j = 1, \dots, k\}.$$

To draw the map $T : X \to X$ into the definition we make the following definitions.

DEFINITION. Let $\alpha = \{A_1, \dots, A_n\}$ be an open cover for a compact metric space X; then for a continuous map $T : X \to X$ we define

$$T^{-1}\alpha = \{T^{-1}A_1, \dots, T^{-1}A_n\}.$$

For $k \geq 1$ we define

$$\vee_{i=0}^{k-1} T^{-i}\alpha = \alpha \vee T^{-1}\alpha \vee \dots \vee T^{-(k-1)}\alpha$$
$$= \{A_{i_0} \cap T^{-1}A_{i_1} \cap \dots^{-(k-1)} A_{i_{k-1}} : 1 \leq i_0, i_1, \dots, i_{k-1} \leq n\}.$$

Given a cover $\alpha = \{A_1, \dots, A_n\}$ we call $\beta \subset \alpha$ a *subcover* if it is still a cover for X (i.e. $X = \cup_{B \in \beta} B$) We can now define the topological entropy of a cover α for X as follows.

DEFINITION. The *topological entropy* of the cover α is defined to be the logarithm $H(\alpha) = \log N(\alpha)$ of the smallest number $N(\alpha)$ of sets that can be used in a subcover of α.

It is useful to illustrate this with a little example.

EXAMPLE. Let $X = \mathbb{R}/\mathbb{Z}$ and consider the open cover

$$\alpha = \left\{ \left(\frac{1}{4}, 1 \right), \left(0, \frac{3}{4} \right), \left(\frac{1}{2}, 1 \right] \cup \left(0, \frac{1}{4} \right), \left(\frac{3}{4}, 1 \right] \cup \left(0, \frac{1}{2} \right) \right\}$$

consisting of four open sets. Here $N(\alpha) = 2$ and so $H(\alpha) = \log 2$.

The next lemma gives some of the basic properties of the topological entropy of covers.

LEMMA 3.1.
  (i) $H(\alpha) \geq 0$.
 (ii) If $\beta \subset \alpha$ is a subcover then $H(\alpha) \geq H(\beta)$.
(iii) If α and β are two finite covers for X then $H(\alpha \vee \beta) \leq H(\alpha) + H(\beta)$.
 (iv) If $T : X \to X$ is continuous and $T(X) = X$ then $H(\alpha) \geq H(T^{-1}\alpha)$.
        If $T : X \to X$ is a homeomorphism then $H(\alpha) = H(T^{-1}\alpha)$.

PROOF. Parts (i) and (ii) follow immediately from the definitions. For (iii) we let $\alpha \supset \alpha' = \{A_1, \ldots, A_n\}$ and $\beta \supset \beta' = \{B_1, \ldots, B_m\}$ be subcovers of α and β of minimal cardinality. We can then write $\alpha' \vee \beta' = \{A_i \cap B_j : 1 \leq i \leq n, 1 \leq j \leq m, A_i \cap B_j \neq \emptyset\}$ which is a subcover of $\alpha \vee \beta$ of cardinality at most $n \times m$. Thus $H(\alpha \vee \beta) \leq \log(nm) = \log(n) + \log(m) = \log H(\alpha) + \log H(\beta)$.

For (iv), we note that if $\alpha \supset \alpha' = \{A_1, \ldots, A_n\}$ is a subset of α of least cardinality then $T^{-1}\alpha' = \{T^{-1}A_1, \ldots, T^{-1}A_n\}$ is a subcover of $T^{-1}\alpha$ of cardinality n. Thus $N(T^{-1}\alpha) \leq n = N(\alpha)$.

If T is surjective, then for a minimal subcover $T^{-1}\alpha' = \{T^{-1}A_1, \ldots, T^{-1}A_m\} \subset T^{-1}\alpha$ we have that $\alpha' = \{A_1, \ldots, A_m\}$ is a subcover of α. Thus $N(T^{-1}\alpha) = m \geq N(\alpha)$. Together these two inequalities give an equality. ∎

We can define the topological entropy of a transformation relative to a cover as follows.

DEFINITION. Let $T : X \to X$ be a continuous map on X; then we define the *topological entropy of T relative to a cover* α by

$$h(T, \alpha) = \limsup_{n \to +\infty} \frac{1}{n} H(\vee_{i=0}^{n-1} T^{-i}\alpha).$$

We can see that $h(T, \alpha) < +\infty$ by the following lemma.

LEMMA 3.2. $\frac{1}{n} H(\vee_{i=0}^{n-1} T^{-i}\alpha) \leq H(\alpha)$ *for* $n \geq 1$.

PROOF. We can write $H(\vee_{i=0}^{n-1} T^{-i}\alpha) \leq \sum_{i=0}^{n-1} H(T^{-i}\alpha) \leq nH(\alpha)$ by parts (iii) and (iv) of Lemma 3.1. ∎

Finally, we can define the topological entropy of a transformation as follows.

DEFINITION. If $T : X \to X$ is a continuous map on a compact metric space then we define the *topological entropy* by

$$h(T) = \sup\{h(T, \alpha) : \alpha \text{ is a finite cover for } X\}.$$

REMARK. In fact the limsup in the definition of the topological entropy of T relative to a cover α can be replaced by a straightforward limit. If we write $a_n = H(\vee_{i=0}^{n-1} T^{-i}\alpha)$ then for $n, m \geq 1$ we have that

$$
\begin{aligned}
a_{n+m} &= H(\vee_{i=0}^{n+m-1} T^{-i}\alpha) \\
&\leq H(\vee_{i=0}^{n-1} T^{-i}\alpha) + H(\vee_{i=n}^{n+m-1} T^{-i}\alpha) \\
&\leq H(\vee_{i=0}^{n-1} T^{-i}\alpha) + H(\vee_{i=0}^{m-1} T^{-i}\alpha) \\
&= a_n + a_m
\end{aligned}
$$

(i.e. this sequence is *subadditive*). In particular, if $a = \inf\{\frac{a_n}{n} : n \geq 1\}$, then we know that $\frac{a_n}{n} \to a$, as $n \to +\infty$.

Although the definition of topological entropy we have given is not particularly convenient for computations, we shall take the opportunity to show that the trivial identity map has zero topological entropy.

TRIVIAL EXAMPLE. Let $T = id : X \to X$ be the identity transformation on a space X. For any cover α we have that $T^{-i}\alpha = \alpha$ and so $\alpha = \vee_{i=0}^{n-1} T^{-i}\alpha$. This means that $H(\alpha) = H(\vee_{i=0}^{n-1} T^{-i}\alpha)$ and so

$$h(T, \alpha) = \limsup_{n \to +\infty} \frac{1}{n} H(\vee_{i=0}^{n-1} T^{-i}\alpha) = \limsup_{n \to +\infty} \frac{1}{n} H(\alpha) = 0.$$

Since this holds for any cover α, we see that $h(T) = 0$.

We shall describe a method of computing topological entropy.

DEFINITION. We call a finite cover α a *generator* for a homeomorphism $T : X \to X$ if $\forall \epsilon > 0 \ \exists N > 0$ the cover $\vee_{n=-N}^{N} T^{-n}\alpha = \{B_1, \ldots, B_m\}$ consists of open sets each of which has diameter at most ϵ, i.e. $\sup_i\{\operatorname{diam}(B_i)\} \leq \epsilon$.

We call a finite cover α a *(strong) generator* for a continuous map $T : X \to X$ if $\forall \epsilon > 0 \ \exists N > 0$ the cover $\vee_{n=0}^{N} T^{-n}\alpha = \{B_1, \ldots, B_m\}$ consists of open sets each of which has diameter at most ϵ, i.e. $\sup_i\{\operatorname{diam}(B_i)\} < \epsilon$.

This brings us to the following useful result.

PROPOSITION 3.3. *If α is a strong generating cover for a continuous map $T : X \to X$ (or a generator for a homeomorphism $T : X \to X$) then $h(T, \alpha) = h(T)$*

PROOF. We prove the result for a strong generator (the proof for a generator being similar). Let β be an arbitrary cover. Let δ be the Lebesgue number for β (i.e. every ball of diameter δ is contained inside some open set in the cover β). For sufficiently large N we get that $\forall A \in \vee_{n=0}^{N-1} T^{-n}\alpha, \exists B \in \beta$ with $A \subset B$, from the definition of α being generating. This means that $N(\vee_{i=0}^{k-1} T^{-i}(\vee_{n=0}^{N-1}))N(\vee_{i=0}^{k-1}T^{-i}\beta)$, for $k \geq 1$.

We claim that $h(T, \alpha) = h(T, \vee_{n=0}^{N}T^{-n}\alpha)$. In fact, the equality

$$H\left(\vee_{i=0}^{k-1}T^{-i}\left(\vee_{n=0}^{N}T^{-n}\alpha\right)\right) = H(\vee_{i=0}^{k+N-1}T^{-i}\alpha), \quad k \geq 1,$$

gives

$$h(T, \vee_{n=0}^{N}T^{-n}\alpha) = \limsup_{k \to +\infty} \frac{1}{k}H\left(\vee_{i=0}^{k-1}T^{-i}\left(\vee_{n=0}^{N}T^{-n}\alpha\right)\right)$$

$$= \limsup_{k \to +\infty} \frac{1}{k}H(\vee_{i=0}^{k+N-1}T^{-i}\alpha) = h(T, \alpha).$$

Altogether we see that $h(T, \alpha) \geq h(T, \beta)$. Since this holds for all open covers β we see that $h(T) = h(T, \alpha)$. ∎

EXAMPLE (FULL SHIFT ON k SYMBOLS). Let $X = \prod_{n \in \mathbb{Z}}\{1, \ldots, k\}$ and $\sigma : X \to X$ be the shift map. We can choose a cover by open sets of the form

$$\alpha = \{[1]_0, \ldots, [k]_0\}, \text{ where } [i]_0 = \{x = (x_n) \in X : x_0 = i\},$$

for $i = 1, \ldots, k$. Observe that

$$\vee_{n=-N}^{N}\sigma^{-i}\alpha = \{[i_{-N}, \ldots, i_0, \ldots, i_N]_{-N}^{N} : i_{-N}, \ldots, i_0, \ldots, i_N \in \{1, \ldots, k\}\}$$

where we write

$$[i_{-N}, \ldots, i_0, \ldots, i_N]_{-N}^{N} = \{x = (x_n) \in X : x_j = i_j \text{ for } -N \leq j \leq N\}.$$

For any $\epsilon > 0$ we can choose $N > 0$ sufficiently large that $\frac{1}{2^N} \leq \epsilon$; then for any $[i_{-N}, \ldots, i_0, \ldots, i_N]_{-N}^{N} \in \vee_{n=-N}^{N}\sigma^{-i}\alpha$ we have that

$$\text{diam}\left([i_{-N}, \ldots, i_0, \ldots, i_N]_{-N}^{N}\right) \leq \epsilon$$

(i.e. the open cover α is a generator). Moreover, each of the sets in $\vee_{n=0}^{N}\sigma^{-i}\alpha$ is disjoint and $\vee_{n=0}^{N}\sigma^{-i}\alpha$ contains k^{N+1} elements. This means that $N(\vee_{n=0}^{N}\sigma^{-i}\alpha) = k^{N+1}$. Thus $h(T) = h(T, \alpha) = \log k$.

3.2 The Perron-Frobenious theorem and subshifts of finite type

We want to consider a subshift of finite type $\sigma : X_A \to X_A$ where A is a $k \times k$ matrix satisfying the following property.

DEFINITION. The matrix A is called *aperiodic* if $\exists N > 0$, $\forall 1 \leq i, j \leq k$, $A(i, j) \geq 1$.

This property is stronger than that of irreducibility introduced in chapter 1. In order to understand the topological entropy we need the following.

LEMMA 3.4 (PERRON-FROBENIUS). *Let A be an aperiodic $k \times k$ matrix; then there exists an positive eigenvalue $\lambda_1 > 0$ such that all other eigenvalues λ_i satisfy $|\lambda_i| < \lambda_1$, $i = 2, \ldots, k$. Moreover, there is a unique positive vector $v = (v_1, \ldots, v_k)$ such that $Av = \lambda_1 v$.*

PROOF. We denote the *positive cone* in $\mathbb{R}^k$ by

$$C = \{x = (x_1, \ldots, x_k) \in \mathbb{R}^k : x_i \geq 0, 1 \leq i \leq k\}.$$

Observe that $A : C \to C$ where

$$A(v_1, \ldots, v_k) = (v_1', \ldots, v_k') = (\sum_{i=1}^{k} A(i, 1)v_i, \ldots, \sum_{i=1}^{k} A(i, k)v_i)$$

and so $v_i' \geq 0$, $i = 1, \ldots, k$ (i.e. $Av \in C$). We denote the *standard simplex* in $\mathbb{R}^k$ by

$$S = \{x = (x_1, \ldots, x_k) \in \mathbb{R}^k : x_i \geq 0, 1 \leq i \leq k, \text{ and } \sum_{i=1}^{k} x_i = 1\} \subset C$$

and we can define $T : S \to S$ by

$$T(v_1, \ldots, v_k) = \left(\frac{\sum_{i=1}^{k} A(i, 1)v_i}{\sum_{i=1}^{k} \sum_{j=1}^{k} A(i, j)v_i}, \ldots, \frac{\sum_{i=1}^{k} A(i, k)v_i}{\sum_{i=1}^{k} \sum_{j=1}^{k} A(i, j)v_i} \right).$$

Observe that for each $n \geq 0$, we can write

$$T^n S = \{\sum_{i=1}^{n} x_i T^n(e_i) : x_i \geq 0, 1 \leq i \leq k, \text{ and } \sum_{i=1}^{k} x_i = 1\}$$

where e_i, $1 \leq i \leq k$, are the standard basis vectors for $\mathbb{R}^n$. Notice that:

(1) because of the assumption that A is aperiodic, we know that for some $N > 0$ we have $A^N(i, j) = 1$, for all $1 \leq i, j \leq k$. In particular, we see that $T^N S \subset \text{int}(S)$ (see figure 3.1); and

(2) $T^N(S)$ is convex.

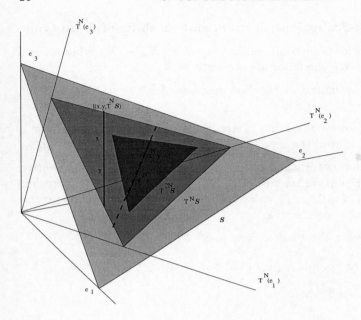

FIGURE 3.1.   The image $T(\mathcal{S}) \subset \text{int}(\mathcal{S})$

We would like to show that $\mathcal{S} \supset T^N(\mathcal{S}) \supset T^{2N}(\mathcal{S}) \supset \ldots \supset T^{nN}(\mathcal{S}) \supset \ldots$ converges to a single point by using the contraction mapping theorem applied to a suitable metric.

Given two points $x, y \in \text{int}(T^N(\mathcal{S}))$ let $l(x, y, T^N(\mathcal{S}))$ be the line segment given by the intersection of the convex set $T^N(\mathcal{S})$ with the unique line in $\mathcal{S}$ through x and y. We can always translate each such segment $l(x, y, T^N(\mathcal{S}))$ to the standard one-dimensional simplex $\Delta = \{(x_1, x_2) \in \mathbb{R}^2 : x_1, x_2 \geq 0, x_1 + x_2 = 1\}$ by a surjective affine transformation $L = L_{x,y,T^N} : l(x, y, T^N(\mathcal{S}) \to \Delta$. Furthermore, we can identify Δ with $\mathbb{R}^+$ by $(x_1, x_2) \mapsto \frac{x_1}{x_2}$.

If we write $L(x) = (x_1, x_2)$ and $L(y) = (y_1, y_2)$ then we can introduce a metric on $\text{int}(T^N\mathcal{S})$ by

$$d(x, y) = \left| \log \left(\frac{x_2 y_1}{x_1 y_2} \right) \right|.$$

We claim that $T^N : T^N(\mathcal{S}) \to T^{2N}(\mathcal{S}) \subset T^N(\mathcal{S})$ is a contraction with respect to this map (i.e. there exists $0 < C < 1$ such that $d(T^N x, T^n y) \leq C d(x, y)$).

We can affinely identify each of the line segments $l(x, y, T^N(\mathcal{S})$ and $l(T^N x, T^N y, T^N(\mathcal{S}))$ with Δ, and hence $\mathbb{R}^+$, as above. The transformation $P : \mathbb{R}^+ \to \mathbb{R}^+$ corresponding to $T^N : l(x, y, T^N(\mathcal{S})) \to l(T^N x, T^N y, T^N(\mathcal{S}))$ is a linear fractional transformation $P(z) = \frac{az+b}{cz+d}$ ($a, b, c, d \in \mathbb{R}^+$). The corresponding metric on $\mathbb{R}$ is given by

$$\theta(z_1, z_2) = |\log\left(\frac{z_1}{z_2}\right)|.$$

To show that T^N is a contraction it suffices to show that each such corresponding P is a contraction, and that the contraction constant $C < 1$ is independent of the choices f, g. Using the usual euclidean distance on $\mathbb{R}^+$ we see that $P'(z) = \frac{ad-bc}{(cz+d)^2}$ and since the ratio of the θ distance relative to the euclidean distance at $z \in \mathbb{R}^+$ is given by $\frac{1}{z}$ we conclude that P is Lipschitz with constant

$$C = \sup_{z \in \mathbb{R}}\left\{z \cdot \left(\frac{ad - bc}{(cz+d)^2}\right)\left(\frac{cz + d}{az + b}\right)\right\}.$$

However, elementary calculus shows that the supremum is realised where $z = \nu := \frac{ad}{bc}$ and thus

$$C = \frac{\nu^{1/4} - \nu^{-1/4}}{\nu^{1/4} + \nu^{-1/4}} < 1.$$

Finally, since $\log(\nu) = \log\left(\frac{ad}{bc}\right) = \theta(P(0), P(+\infty))$ we conclude from the fact that $T^N(T^N\mathcal{S}) \subset \mathrm{int}(T^N\mathcal{S})$ that ν is uniformly bounded over all choices $l(x, y, T^N\mathcal{S})$. In particular, we can choose a value $C < 1$ valid for all such choices.

This completes the proof of the lemma. ∎

This lemma has immediate applications to the topological entropy of mixing subshifts of finite type.

PROPOSITION 3.5. *Let A be an aperiodic $k \times k$ matrix with entries 0 or 1. Let $\sigma : X_A \to X_A$ be the associated subshift of finite type and let λ_1 be the maximal positive eigenvalue; then $h(\sigma) = \log \lambda_1$.*

PROOF. We can choose a cover for X_A by the disjoint open sets of the form $\alpha = \{[1]_0, \dots, [k]_0\}$ where $[i]_0 = \{x = (x_n) \in X_A : x_0 = i\}, i = 1, \dots, k$. We see that α is a generating cover.

Since all of the open sets in the cover $\vee_{n=0}^{N-1}\sigma^{-i}\alpha$ are disjoint we see that

$$H\left(\vee_{n=-N}^{N}\sigma^{-i}\alpha\right) = \log\mathrm{Card}\left(\vee_{n=0}^{N-1}\sigma^{-i}\alpha\right)$$

$$= \log\left(\sum_{i,j}A^N(i,j)\right).$$

From elementary linear algebra there are an invertible matrix U and a matrix

$$D = \begin{pmatrix} \lambda_1 & 0 & \dots & 0 \\ 0 & B_2 & \dots & 0 \\ \vdots & \vdots & \ddots & \vdots \\ 0 & 0 & \dots & B_l \end{pmatrix},$$

with Jordan block matrices $B_1, \ldots, B_l$ $(1 \leq l \leq k)$, such that $A = UDU^{-1}$.
In particular, we can write

$$\sum_{i,j} A^N(i,j) = \sum_{i,j} (UD^NU^{-1})(i,j)$$

$$= C\lambda_1^N + E(n)$$

where

(i) $C = \left(\sum_{i,j} U(i,1)U^{-1}(1,j) \right)$, and

(ii) $\lim_{N \to +\infty} \frac{|E(N)|}{\lambda_1^N} = 0$.

Thus we see that

$$\lim_{N \to +\infty} \frac{1}{N} H \left(\vee_{i=0}^{N-1} \sigma^{-i} \alpha \right) = \log \lambda_1.$$

$\blacksquare$

3.3 Other definitions and examples

There are equivalent definitions of topological entropy which, in some
examples, are easier to apply. In this section we shall describe two such
definitions.

DEFINITION. Let $T : X \to X$ be a continuous map on a compact metric
space.

(1) For $n \geq 1$ and $\epsilon > 0$ we call a finite set $S \subset X$ an (n, ϵ)-separated set
   if for distinct points $x, y \in S$ we have that $d(T^i x, T^i y) \geq \epsilon$ for some
   $0 \leq i \leq n - 1$.
   Let $s(n, \epsilon) \in \mathbb{N}$ denote the maximal cardinality of any (n, ϵ)-
   separated set.

(2) For $n \geq 1$ and $\epsilon > 0$ we call a finite set $R \subset X$ an (n, ϵ)-spanning set
   if $\forall x \in X$, $\exists y \in R$ such that $d(T^i x, T^i y) \leq \epsilon$ for all $i = 0, \ldots, n - 1$.
   Let $r(n, \epsilon) \in \mathbb{N}$ denote the least cardinality of any (n, ϵ)-spanning
   set.

LEMMA 3.6.

(i) For $\epsilon > \epsilon'$ we have that $s(n, \epsilon') > s(n, \epsilon)$ and $r(n, \epsilon') > r(n, \epsilon)$. In
   particular,

$$\limsup_{n \to +\infty} \frac{1}{n} \log \left(s(n, \epsilon') \right) \geq \limsup_{n \to +\infty} \frac{1}{n} \log \left(s(n, \epsilon) \right)$$

and

$$\limsup_{n \to +\infty} \frac{1}{n} \log \left(r(n, \epsilon') \right) \geq \limsup_{n \to +\infty} \frac{1}{n} \log \left(r(n, \epsilon) \right).$$

(ii) For $\epsilon > 0$ and $n \geq 1$ we have that

$$r(n, \epsilon) \leq s(n, \epsilon) \quad and \quad s(n, 2\epsilon) \leq r(n, \epsilon).$$

PROOF. Part (i) follows directly from the definitions.

For part (ii) we first observe from the definitions that $r(n,\epsilon) \geq s(n,\epsilon)$ since an (n,ϵ)-separated set S of maximum cardinality $r(n,\epsilon)$ must also be an (n,ϵ)-spanning set.

If R is an (n,ϵ)-spanning set of least cardinality $r(n,\epsilon)$ then we know that $X = \cup_{x \in R} D(x,n,\epsilon)$ where

$$D(x,n,\epsilon) = \{y \in X : d(T^i x, T^i y) \leq \epsilon, \quad i = 1, \ldots, n\}.$$

If S is an $(n,2\epsilon)$-separated set, then for each $y \in S$ we can choose a *distinct* open set $D(x,n,\epsilon) \ni y$, with $x \in R$ (since if $y, y' \in S$ and $x \in R$ with $y, y' \in D(x,n,\epsilon)$ then $d(T^i y, T^i y') \leq d(T^i y, T^i x) + d(T^i x, T^i y') \leq \epsilon + \epsilon = 2\epsilon$ which, since S is $(n,2\epsilon)$-separated, implies that $y = y'$). Thus $\mathrm{Card}(S) \leq r(n,\epsilon)$ for all such sets S, i.e. $s(n,2\epsilon) \leq r(n,\epsilon)$. ∎

The values $r(n,\epsilon)$ and $s(n,\epsilon)$ tend to be more tractable than, say, $N(\vee_{i=0}^{n-1} T^{-i}\alpha)$. These quantities are compared by the following lemma.

LEMMA 3.7.

(i) *If α is a finite open cover for X with Lebesgue number δ then*

$$N\left(\vee_{i=0}^{n-1} T^{-i}\alpha\right) \leq r(n,\delta)$$

for all $n \geq 1$.

(ii) *If $\epsilon > 0$ and $\gamma = \{B_1, \ldots, B_k\}$ is an open cover with $\max_{1 \leq i \leq k}$ diam $(B_i) < \epsilon$, then*

$$s(n,\epsilon) \leq N\left(\vee_{i=0}^{n-1} T^{-i}\gamma\right)$$

PROOF. (i) If R is an (n,δ)-spanning set of maximum cardinality $r(n,\delta)$ then $X = \cup_{x \in R} D(x,n,\delta)$. However, for each $x \in R$ we have $B(T^j x, \delta) \subset A_{i_j}$, where $A_{i_j} \in \alpha$, for $0 \leq j \leq n-1$, i.e. $D(x,n,\delta) \subset A_{i_0} \cap T^{-1} A_{i_1} \cap \cdots \cap T^{-(n-1)} A_{i_{n-1}} \in \vee_{i=0}^{n-1} T^{-i}\alpha$. In particular, since these sets form a subcover for X we see that $N\left(\vee_{i=0}^{n-1} T^{-i}\alpha\right) \leq r(n,\delta)$.

(ii) Let S be a (n,ϵ)-separated set of cardinality $s(n,\epsilon)$. Each point $x \in S$ must lie in a different element of $\vee_{i=0}^{n-1} T^{-i}\gamma$ (since for $x, y \in B_{i_0} \cap T^{-1} B_{i_1} \cap \ldots \cap T^{-(n-1)} B_{i_{n-1}} \in \vee_{i=0}^{n-1} T^{-i}\gamma$ we see that $d(T^r x, T^r y) \leq$ diam $(B_{i_r}) < \epsilon$). In particular, $s(n,\epsilon) \leq N\left(\vee_{i=0}^{n-1} T^{-i}\gamma\right)$. ∎

This gives us the following equivalent definitions of topological entropy

PROPOSITION 3.8. *The topological entropy of a continuous map* $T : X \to$ *X on a compact metric space is given by*

$$h(T) = \lim_{\epsilon \to 0} \limsup_{n \to +\infty} \frac{1}{n} \log\left(r(n, \epsilon)\right)$$

and

$$h(T) = \lim_{\epsilon \to 0} \limsup_{n \to +\infty} \frac{1}{n} \log\left(s(n, \epsilon)\right).$$

PROOF. Since $h(T) = \sup_\alpha h(T, \alpha)$ we can choose for any $\eta > 0$ a finite cover α such that

$$h(T, \alpha) + \eta \geq h(T) \geq h(T, \alpha).$$

Let $\delta > 0$ be the Lebesgue number for the open cover α.

By Lemma 3.6 we see that the two limits are the same, i.e.

$$\lim_{\epsilon \to 0} \limsup_{n \to +\infty} \frac{1}{n} \log\left(r(n, \epsilon)\right) = \lim_{\epsilon \to 0} \limsup_{n \to +\infty} \frac{1}{n} \log\left(s(n, \epsilon)\right). \tag{3.1}$$

Observe that we have the following inequalities for $n \geq 1$:

$$\begin{aligned}
&\lim_{\epsilon \to 0} \limsup_{n \to +\infty} \frac{1}{n} \log\left(r(n, \epsilon)\right) \\
&\geq \limsup_{n \to +\infty} \frac{1}{n} \log\left(r(n, \delta)\right) \qquad \text{(by Lemma 3.6 (i))} \\
&\geq \limsup_{n \to +\infty} \frac{1}{n} \log N\left(\vee_{i=0}^{n-1} T^{-i}\alpha\right) \qquad \text{(by Lemma 3.7 (i))} \\
&= h(T, \alpha) \geq h(T) - \eta.
\end{aligned} \tag{3.2}$$

For any $\epsilon > 0$ we can choose an open cover $\beta = \{B_1, \dots, B_k\}$ for X with $\max_{1 \leq i \leq k}\{\text{diam}(B_i)\} < \epsilon$. This gives the following inequalities:

$$\begin{aligned}
&\limsup_{n \to +\infty} \frac{1}{n} \log\left(s(n, \epsilon)\right) \\
&\leq \limsup_{n \to +\infty} \frac{1}{n} \log H\left(\vee_{i=0}^{n-1} T^{-i}\beta\right) \quad \text{(by Lemma 3.7 (ii))} \\
&\leq h(T, \beta) \leq h(T).
\end{aligned}$$

Letting $\epsilon \to 0$ now gives that

$$\lim_{\epsilon \to 0} \limsup_{n \to +\infty} \frac{1}{n} \log\left(s(n, \epsilon)\right) \leq h(T). \tag{3.3}$$

Comparing the estimates (3.1),(3.2) and (3.3) (and recalling that $\eta > 0$ can be chosen arbitrarily small) completes the proof.

∎

EXAMPLE (ROTATIONS). Let $T : \mathbb{R}/\mathbb{Z} \to \mathbb{R}/\mathbb{Z}$ by $Tx = x + \alpha$ (mod 1). We claim that $h(T) = 0$.

Observe that for $\epsilon > 0$ and $n = 1$ we choose a finite cover for $[0,1]$ consisting of intervals $\left(x_i - \frac{\epsilon}{2}, x_i + \frac{\epsilon}{2}\right)$, $i = 1, \ldots, N$, say. Thus $R = \{x_i : i = 1, \ldots, N\}$ is a $(1, \epsilon)$-spanning set and $r(1, \epsilon) \leq N$.

We claim that R is also an (n, ϵ)-spanning set for each $n \geq 1$. For any $x \in [0, 1]$ we can choose $d(x, x_i) \leq \epsilon$ then $d(T^r x_i, T^r x) = d(x, x_i) < \epsilon$, for $r = 0, \ldots, N - 1$. In particular, $r(n, \epsilon) \leq N$. Thus we see that

$$0 \leq h(T) = \lim_{\epsilon \to 0} \left(\limsup_{n \to +\infty} \frac{1}{n} \log r(n, \epsilon) \right)$$
$$\leq \lim_{\epsilon \to 0} \left(\limsup_{n \to +\infty} \frac{1}{n} \log N \right)$$
$$= 0.$$

REMARK. This proof works equally well for any isometry $T : X \to X$ on a compact metric space.

EXAMPLE (DOUBLING MAP). Consider the map $T : \mathbb{R}/\mathbb{Z} \to \mathbb{R}/\mathbb{Z}$ defined by $Tx = 2x$ (mod 1).

For $k \geq 1$ we can define the set

$$F_k = \{\frac{m}{2^k} : m = 0, \ldots, 2^k - 1\}$$

(whose cardinality is 2^k). For $\epsilon > 0$ choose $k \geq 1$ such that $\frac{1}{2^k} \leq \epsilon < \frac{1}{2^{k-1}}$.

For $n \geq 1$ we claim that F_{n+k-2} is an (n, ϵ)-separated set. This is clear since for distinct points $\frac{m_1}{2^{k+n-2}}, \frac{m_2}{2^{k+n-2}} \in F_{n+k}$ we have that $d(T^{n-1}(\frac{m_1}{2^{k+n}}), T^{n-1}(\frac{m_2}{2^{k+n}})) = d(\frac{m_1}{2^{k-1}}, \frac{m_1}{2^{k-1}}) \geq \frac{1}{2^{k-1}} \geq \epsilon$. Thus $s(n, \epsilon) \geq 2^{k+n-2}$.

For $n \geq 1$ we claim that F_{n+k} is an (n, ϵ)-spanning set. For any $x \in \mathbb{R}/\mathbb{Z}$ we can choose $\frac{m}{2^{k+n}} \in F_{n+k}$ with $d(x, \frac{m}{2^{k+n}}) \leq \frac{1}{2^{k+n}}$. Then for $r = 0, 1, \ldots, n - 1$ we have that $d(T^r \frac{m}{2^{k+n}}, T^r x) \leq \frac{1}{2^{k+1}} \leq \epsilon$. Thus $r(n, \epsilon) \leq 2^{k+n-1}$.

TWe now know that

$$h(T) = \lim_{\epsilon \to 0} \left(\limsup_{n \to +\infty} \frac{1}{n} \log s(n, \epsilon) \right) \geq \log 2$$

and

$$h(T) = \lim_{\epsilon \to 0} \left(\limsup_{n \to +\infty} \frac{1}{n} \log r(n, \epsilon) \right) \leq \log 2.$$

Together these inequalities show that $h(T) = \log 2$.

Finally, we give a result that will be useful later on.

COROLLARY 3.8.1 (ABRAMOV'S THEOREM). *Let $T : X \to X$ be a continuous map on a compact metric space X; then for any $m \geq 1$ we have* $h(T^m) = mh(T)$.

PROOF. Let $T : X \to X$ be a continuous map on a compact metric space X. For $m \geq 1$, let $s_{T^m}(n, \epsilon)$ denote the maximal cardinality of an (n, ϵ)-separated set for T^m and let $r_{T^m}(n, \epsilon)$ denote the least cardinality of an (n, ϵ)-spanning set for T^m. We shall show that $s_{T^m}(n, \epsilon) \geq s_T(nm, \epsilon)$ and $r_{T^m}(n, \epsilon) \leq_T (nm, \epsilon)$. Let S be an (n, ϵ)-separated set for T^m of cardinality $s_{T^m}(n, \epsilon)$. Then $\forall$ (distinct) $x, x' \in X$, $\exists 0 \leq i \leq n-1$ with $d(T^{mi}x, T^{mi}x') \geq \epsilon$. In particular, S is also an (nm, ϵ)-separated set for T. Thus $s_{T^m}(n, \epsilon)$ is a *lower* bound on the maximal cardinality of (n, ϵ)-separated sets for T:
$$s_{T^m}(n, \epsilon) \geq s_T(nm, \epsilon)$$

Similarly, let R be an (nm, ϵ)-spanning set for T of maximal cardinality $r_T(nm, \epsilon)$; then $\forall x \in X, \exists z \in R$ such that $d(T^i x, T^i z) < \epsilon$ for $0 \leq i \leq nm-1$. In particular, R is also an (n, ϵ)-spanning set for T^n and so $r_{T^m}(n, \epsilon) \leq r_T(nm, \epsilon)$.

Finally, we have that

$$h(T^m) = \lim_{\epsilon \to 0} \lim_{n \to +\infty} \frac{1}{n} r_{T^m}(n, \epsilon) \leq \lim_{\epsilon \to 0} \lim_{n \to +\infty} \frac{1}{n} r_T(nm, \epsilon) = mh(T) \text{ and}$$

$$h(T^m) = \lim_{\epsilon \to 0} \lim_{n \to +\infty} \frac{1}{n} s_{T^m}(n, \epsilon) \geq \lim_{\epsilon \to 0} \lim_{n \to +\infty} \frac{1}{n} s_T(nm, \epsilon) = mh(T).$$

which completes the proof. ∎

3.4 Conjugacy

We begin with the following definition.

DEFINITION. Two homeomorphisms (or continuous maps) $T_1 : X_1 \to X_1$ and $T_2 : X_2 \to X_2$ are *conjugate* if there exists a homeomorphism $h : X_1 \to X_2$ such that $h \circ T_1 = T_2 \circ h$.

REMARKS.

(i) The existence of such a homeomorphism requires X_1 and X_2 to be homeomorphic (before any consideration is made of the maps T_1 and T_2).

(ii) The condition $T_2 \circ h = h \circ T_1$ means that h respects the orbits of T_1 and T_2, i.e. for any $x \in X_1$ and $n \geq 0$ (or $n \in \mathbb{Z}$ for homeomorphisms) we have that $h(x) \in X_2$ and $h(T_1^n x) = T_2^n(hx)$

(iii) Conjugacy is an equivalence relation on the set of all homeomorphisms on compact metric spaces.

The following is easy to prove from the definitions.

PROPOSITION 3.9. *Assume that $T_1 : X_1 \to X_1$ and $T_2 : X_2 \to X_2$ are conjugate maps then;*

(1) $T_1 : X_1 \to X_1$ *is transitive if and only if* $T_2 : X_2 \to X_2$ *is transitive,*

(2) $T_1 : X_1 \to X_1$ *is minimal if and only if* $T_2 : X_2 \to X_2$ *is minimal.*

PROOF. This is immediate from the definitions and the easy observation that the homeomorphic image of a dense set is dense.

∎

DEFINITION. Denote by $\text{Fix}(T^n)$ the set of fixed points for the nth iterate $T^n : X \to X$ of a map $T : X \to X$ (also called periodic points). We let $\text{Card}\,(\text{Fix}(T^n))$ denote the cardinality of this set (which might be empty, finite or infinite).

Remark (ii) above now gives us that the number of fixed points for T^n gives us a simple set of invariants for conjugacy.

PROPOSITION 3.10. *Conjugate homeomorphisms T_1 and T_2 have the property that $h\,(\text{Fix}(T_1^n)) = \text{Fix}(T_2^n)$ and $\text{Card}\,(\text{Fix}(T_1^n)) = \text{Card}\,(\text{Fix}(T_2^n))$, for $n \geq 1$.*

PROOF. This is immediate from Remark (ii).

∎

EXAMPLES.

(i) Let $T : \mathbb{R}/\mathbb{Z} \to \mathbb{R}/\mathbb{Z}$ be the homeomorphism given by $Tx = x + \alpha \pmod 1$.

If α is irrational then there are no fixed points or periodic points. To see this, assume that $T^n x = x$; then $x + n\alpha = x + m$ with $n, m \in \mathbb{Z}$. Then $\alpha = \frac{m}{n}$, contradicting the fact that α is irrational.

If α is rational ($\alpha = \frac{p}{q}$, where $p, q \geq 1$ are co-prime, i.e. they have no common divisors) then $\text{Fix}(T^n) = \mathbb{R}/\mathbb{Z}$ if n is a multiple of q and the empty set otherwise.

(ii) Let $\sigma : X \to X$ be a subshift of finite type described by a $k \times k$ matrix with entries either 0 or 1. A fixed point will be of the form

$$x = (\ldots, x_0, x_0, x_0, x_0, x_0, \ldots)$$

where $A(x_0, x_0) = 1$. We see that $\text{Card}(\text{Fix}(T)) = \text{trace}(A)$ and similarly $\text{Card}(\text{Fix}(T^n)) = \text{trace}(A^n)$.

DEFINITION. Let $T_1 : X_1 \to X_1$ and $T_2 : X_2 \to X_2$ be continuous maps on compact metric spaces. A continuous map $h : X_1 \to X_2$ is called a *semi-conjugacy* if $h \circ T_1 = T_2 \circ h$ and $h(X_1) = X_2$.

If $h : X_1 \to X_2$ is homeomorphism with $h \circ T_1 = T_2 \circ h$ is conjugacy.

The following relates the entropies of semi-conjugate and conjugate maps.

PROPOSITION 3.11. *If $T_1 : X_1 \to X_1$ is semi-conjugate to $T_2 : X_2 \to X_2$ then $h(T_1) \geq h(T_2)$. If $T_1 : X_1 \to X_1$ and $T_2 : X_2 \to X_2$ are conjugate then $h(T_1) = h(T_2)$*

PROOF. Let α be an open cover for X_2. Then for any $n \geq 1$ we have that

$$
\begin{aligned}
H\left(\vee_{i=0}^{n-1} T_2^{-i}\alpha\right) &= H\left(h^{-1}\left(\vee_{i=0}^{n-1} T_2^{-i}\alpha\right)\right) \\
&= H\left(\vee_{i=0}^{n-1}\left(h^{-1}T_2^{-i}\alpha\right)\right) \\
&= H\left(\vee_{i=0}^{n-1}T_1^{-i}\left(h^{-1}\alpha\right)\right).
\end{aligned}
$$

In particular,

$$
\begin{aligned}
h(T_2, \alpha) &= \limsup_{n \to +\infty} \frac{1}{n} H\left(\vee_{i=0}^{n-1} T_2^{-i}\alpha\right) \\
&= \limsup_{n \to +\infty} \frac{1}{n} H\left(\vee_{i=0}^{n-1}T_1^{-i}\left(h^{-1}\alpha\right)\right) \\
&= h(T_1, h^{-1}\alpha).
\end{aligned}
$$

This gives us that

$$
h(T_2) = \sup_{\alpha} h(T_2, \alpha) = \sup_{h^{-1}\alpha} h(T_1, h^{-1}\alpha) \leq \sup_{\beta} h(T_1, \beta) = h(T_1).
$$

If $h : X_1 \to X_2$ is a homeomorphism then we can interchange $T_1 : X_1 \to X_1$ and $T : X_2 \to X_2$ in the above argument (and replace $h : X_1 \to X_2$ by the homeomorphism $h^{-1} : X_2 \to X_1$) to get $h(T_1) \leq h(T_2)$. ∎

3.5 Comments and references

A more detailed treatment of topological entropy and its properties can be found in Walters' book [3] (or in Bowen's notes [2]). The original reference for this material is [1].

References

1. R. Adler, A. Konheim and M. McAndrew, *Topological entropy*, Trans. Amer. Math. Soc. **114** (1965), 61-85.
2. R. Bowen, *Equilibrium States and Ergodic Theory of Anosov Diffeomorphisms*, Lecture Notes in mathematics, 470, Springer, Berlin, 1975.
3. P. Walters, *An Introduction to Ergodic Theory*, Graduate Text in Mathematics, 79, Springer, Berlin, 1982.

CHAPTER 4

INTERVAL MAPS

In this chapter we shall concentrate on the special case of continuous maps on the closed interval $I = [0, 1]$. This level of specialization allows us to prove some particularly striking results on periodic points and topological entropy.

4.1 Fixed points and periodic points

Let $T : I \to I$ be a continuous map of the interval $I = [0, 1]$ to itself. Recall that a fixed point $x \in I$ satisfies $Tx = x$ and that a periodic point (of period n) satisfies $T^n x = x$. We say that x has *prime period* n if n is the smallest positive integer with this property (i.e. $T^k x \neq x$ for $k = 1, \ldots, n - 1$).

For interval maps a very simple visualization of fixed points exists. We can draw the graph $\mathcal{G}_T$ of $T : I \to I$ and the diagonal $\mathcal{D} = \{(x, x) : x \in I\}$.

LEMMA 4.1. *The fixed points $Tx = x$ occur at the intersection points $(x, x) \in \mathcal{G}_T \cap \mathcal{D}$ (see figure 4.1).*

Similarly, if for $n \geq 2$ we look for intersections of the graph $\mathcal{G}_{T^n}$ (of n-compositions $T^n : I \to I$) with the diagonal $\mathcal{D}$ then the intersection points $(x, x) \in \mathcal{G}_{T^n} \cap \mathcal{D}$ are periodic points of period n.

LEMMA 4.2. *Assume that we have an interval $J \subset I$ with $T(J) \supset J$; then there exists a fixed point $Tx = x \in J$.*

PROOF. We see that showing that there exists $Tx = x \in J$ is equivalent to showing the restriction of the graph $\mathcal{G}_T$ to the portion above J intersects the diagonal $\mathcal{D}$. This is obvious by the intermediate value theorem and figure 4.1.

∎

The following simple lemma will prove useful.

LEMMA 4.3. *If $T : I \to I$ is a continuous map and $J_1, J_2 \subset I$ are (closed) sub-intervals with $T(J_1) \supset J_2$ then we choose a sub-interval $J_0 \subset J_1$ with $T(J_0) = J_2$.*

PROOF. Let $J_1 = [a, b]$ and introduce the disjoint closed sets $A = \{x \in J_1 : T(x) = a\}$ and $B = \{y \in J_1 : T(y) = b\}$. Choose $a' \in A$, $b' \in B$

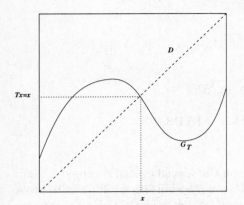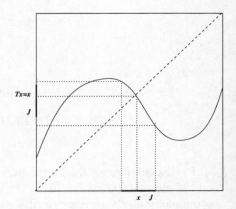

FIGURE 4.1.   Characterizing fixed points (Lemma 4.1) and the existence of fixed points in $J \subset T(J)$ (Lemma 4.2)

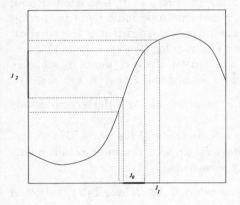

FIGURE 4.2.   Choosing a subinterval $J_0 \subset J_1$ for which $T(J_0) = J_2$

such that $|a' - b'| = \inf\{|x - y| : x \in A, y \in B\}$; then with $J_0 = [a', b']$ or
$J_0 = [b', a']$ the results follows.   ∎

Our first theorem shows the significance of having a periodic point of prime
period 3.

THEOREM 4.4. *Let* $T : I \to I$ *be a continuous map and suppose there
exists a periodic point* x *of prime period* 3. *Then for all* $n \geq 1$ *there exists a
periodic point of prime period* n *(i.e.* $\forall n \geq 1, \exists z \in I$ *with* $T^n z = z$).

PROOF. We shall do the simpler case $n = 1$ and the trickier case $n \geq 2$
separately.

(I) Existence of a fixed point (i.e. $n = 1$). Let $\{x, Tx, T^2x\}$ be the three distinct points in the orbit of x. Let us assume (for simplicity) that $x < Tx < T^2x$. (The five other permutations are easy to derive from this case either by replacing x by Tx or T^2x, or by reversing the horizontal axis of the graph of T.)

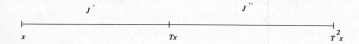

FIGURE 4.3. Intervals J' and J'' defined with endpoints containing the orbit of x

Let $J = [x, T^2x]$; then we can write $J = J' \cup J''$ with $J' = [x, Tx]$ and $J'' = [Tx, T^2x]$ (see figure 4.3). With these choices we have

(a) $T(J') \supset J''$, and
(b) $T(J'') \supset J$

since the endpoints of the intervals J' and J'' are mapped to the endpoints of J' and J, respectively, and the continuous image of an interval is again an interval.

The existence of a fixed point in J'' now follows immediately from Lemma 4.2 and (b) since $T(J'') \supset J \supset J''$.

(II) Existence of a point of period $n \geq 2$. Since by the hypothesis of the theorem we already have a point x of prime period 3 we shall assume henceforth that $n \neq 3$.

SUBLEMMA 4.4.1. *There exists a nested sequence of intervals*

$$J'' = I_0 \supset I_1 \supset I_2 \supset \ldots \supset I_{n-2}$$

with the following properties:

(i) $I_k = T(I_{k+1})$ *for* $k = 0, \ldots, n - 3$;
(ii) $T^{n-1}(I_{n-2}) = J'$; *and*
(iii) $T^n(I_{n-2}) \supset J''$.

To see that Sublemma 4.4.1 implies Theorem 4.4 we first observe that by part (iii) we have that $T^n(I_{n-1}) \supset J'' \supset I_{n-1}$ and so applying Lemma 4.2 (with T replaced by the n-fold composition T^n) shows the existence of a fixed point $z = T^n z \in I_{n-1}$ for T^n (i.e. z is a point of period n for $T : I \to I$). However, we still have to show that this is a periodic point of *prime* period n. We see from Sublemma 4.4.1 that

$z, Tz, T^2z, \ldots, T^{n-2}z \in J''$ (by part (i) since $T^iz \in T^i(I_{n-1}) = I_{n-i-2}$),

$T^{n-1}z \in J'$ (by part (ii)).

$$(4.1)$$

To proceed, we want to eliminate the possibility that $z = Tx(\in J' \cap J'')$. Assume for a contradiction that $z = Tx$, then we immediately have $T^3 z = z$ (since $T^3 x = x$) and z is a periodic point of prime period 3 (since the same is true of x). However, this contradicts our assumption that $n \neq 3$.

In particular, this means that in (4.1) we can "improve" the second conclusion to $T^{n-1} z \notin J'$ We are now in a position to see that n is the prime period of z. If this were not the case, then $T^k z = z$ for some $1 \leq k \leq n - 1$ (which must divide n). But this would mean, in particular, that $T^{k-1} z = T^{n-1} z \notin J'$ which is inconsistent with the first line of (4.1)

The only thing that remains in order to complete the proof of Theorem 4.4 is to prove Sublemma 4.4.1 .

PROOF OF SUBLEMMA 4.4.1.

(i) We know from (b) that $T(J'') \supset J \supset J''$ and so by Lemma 4.3 we can choose $I_1 \subset J''$ with $T(I_1) = J''$. Similarly, since $T(I_1) = J'' \supset I_1$ we can apply Lemma 4.3 again to choose $I_2 \subset I_1$ with $T(I_2) = I_1$.

Proceeding inductively, we can construct a sequence $J'' \supset I_1 \supset I_2 \supset \ldots \supset I_{n-2}$ with $T(I_k) = I_{k-1}$ for $k = 1, 2, \ldots, n-2$ (which, in particular, implies that $T^k(I_k) = J''$ for $k = 1, 2, \ldots n-2$).

(ii) To construct I_{n-1}, observe that $T^{n-1}(I_{n-2}) = T(J'') \supset J'$ by (b). Applying Lemma 4.3 we can find $I_{n-1} \subset I_{n-2}$ with $T^{n-1}(I_{n-1}) = J'$.

(iii) Finally, we observe that $T^n(I_{n-1}) = T(J') \supset J''$ (by (a)).

This completes the proof of the sublemma (and consequently of Theorem 4.4).

∎

This result that a point of prime period 3 implies points of all possible prime periods is a special case of a more general result due to Sharkovski.

We can introduce a new ordering on the natural numbers $\mathbb{N}$ by

$$3 \prec 5 \prec 7 \prec 9 \prec 11 \prec \ldots \prec 2m+1 \prec \ldots$$

$$\ldots \prec 6 \prec 10 \prec 14 \prec 18 \prec 22 \prec \ldots \prec 2(2m+1) \prec \ldots$$

$$\ldots \prec 12 \prec 20 \prec 28 \prec 36 \prec 44 \prec \ldots \prec 4(2m+1) \prec \ldots$$

$$\ldots \quad\quad \ldots \quad\quad \ldots$$

$$\ldots \prec 2^r \cdot 3 \prec 2^r \cdot 5 \prec 2^r \cdot 7 \prec 2^r \cdot 9 \prec 2^r \cdot 11 \prec \ldots \prec 2^r(2m+1) \prec \ldots$$

$$\ldots \quad\quad \ldots \quad\quad \ldots$$

$$\ldots \prec 2^{r+1} \prec 2^r \prec 2^{r-1} \prec \ldots \prec 16 \prec 8 \prec 4 \prec 2 \prec 1$$

This ordering is clearly somewhat different from the usual ordering on the natural numbers. For example, the ordering of the first dozen natural numbers becomes $3 \prec 5 \prec 7 \prec 9 \prec 11 \prec 6 \prec 10 \prec 12 \prec 8 \prec 4 \prec 2 \prec 1$.

THEOREM (SHARKOVSKI). *Let $T : I \to I$ be a continuous map and assume that T has a point of prime period n. Then for each $m > n$ (with respect to the above ordering) there exist periodic points of prime period m.*

The proof of Sharkovski's theorem runs along similar lines to that of Theorem 4.4.

4.2 Topological entropy of interval maps

Given a continuous map $T : I \to I$ we want to express the topological entropy in terms of the growth of the number of monotone intervals for the n th iterate T^n.

DEFINITION. We let $\mathcal{N}(T)$ denote the number of intervals of monotonicity for the map $T : I \to I$ (i.e. the number of disjoint maximal subintervals $I_1, I_2, \ldots, I_k$ for which each restriction $T : I_j \to I$ is strictly monotone).

LEMMA 4.5. *For continuous maps $S_1 : I \to I$ and $S_2 : I \to I$ we have that $\mathcal{N}(S_1 \circ S_2) \leq \mathcal{N}(S_2).\mathcal{N}(S_1)$. In particular, for a continuous map $T : I \to I$*

(i) $\mathcal{N}(T^{n+m}) \leq \mathcal{N}(T^n)\mathcal{N}(T^m)$ *for $n, m \geq 1$,*

(ii) *the sequence $\mathcal{N}(T^n)$ is monotone increasing.*

PROOF. Assume that $\{I_i\}_{i=1}^n$ are disjoint intervals of monotonicity for $S_1 : I \to I$ and $\{J_j\}_{i=1}^m$ are disjoint intervals of monotonicity for $S_2 : I \to I$. The intervals of monotonicity for $S_1 \circ S_2 : I \to I$ take the form $S_2^{-1}(S_2 J_j \cap I_i)$. Thus $\mathcal{N}(S_1 \circ S_2) \leq \mathcal{N}(S_2) \cdot \text{Card}\{i : S_2 I \cap I_i \neq \emptyset\} \leq N(S_1) \cdot N(S_2)$. ∎

REMARK. If we were to also consider C^1 maps $T : I \to I$ then the endpoints of the intervals of monotonicity would be the critical points $\{c : T'(c) = 0\}$. The above results become even more transparent using the chain rule $(T^n)'(x) = \prod_{i=1}^{n-1} T'(T^i x)$.

Lemma 4.5 (i) shows that the sequence $\log N(T^n)$ is subadditive and so $\frac{1}{n} \log \mathcal{N}(T^n)$ converges (to $\inf \left\{ \frac{1}{n} \log \mathcal{N}(T^n) : n \geq 1 \right\}$).

THEOREM 4.6. *Assume that $T : I \to I$ is a continuous map with $\mathcal{N}(T) < +\infty$; then*

$$h(T) = \lim_{n \to +\infty} \frac{1}{n} \log \mathcal{N}(T^n).$$

PROOF. For each $n \geq 1$ let $E_n = \{x_1, x_2, \ldots, x_N\}$ be a maximal (n, ϵ)-separated set for $T : I \to I$ (i.e. $N = s(n, \epsilon)$). By definition, for each $1 \leq i \leq N$ there exists some $1 \leq r_i \leq n$ with $|T^{r_i}(x_i) - T^{r_i}(x_{i+1})| > \epsilon$. In particular, we can choose $1 \leq r \leq n$ with $|T^r(x_{i_j}) - T^r(x_{i_j+1})| > \epsilon$ for a subset $\{x_{i_0} < x_{i_1} < \ldots < x_{i_m}\} \subset E_n$ with cardinality $m \geq \frac{N}{n}$.

If $C = \sup_{x \in I} |T(x)|$ then this property implies there are at least $\frac{m}{C\epsilon}$ intervals of monotonicity for T^r, i.e. $\mathcal{N}(T^r) \geq \frac{m}{C\epsilon} \geq \frac{N}{Cn\epsilon}$. Thus

$$\lim_{n \to +\infty} \frac{1}{n} \log \mathcal{N}(T^n) \geq \lim_{n \to +\infty} \frac{1}{n} \left(\log s(n, \epsilon) - (\log C + \log n + \log \epsilon) \right)$$

$$= \lim_{n \to +\infty} \frac{1}{n} \log s(n, \epsilon) \tag{4.2}$$

$$= h(T).$$

The opposite inequality is slightly more complicated. We begin with some preliminary estimates. We can fix $m \geq 1$ and consider $S := T^m : I \to I$. Observe that

$$m\left(\lim_{n \to +\infty} \frac{1}{n} \log \mathcal{N}(T^n)\right) = m\left(\liminf_{n \geq 1} \frac{1}{n} \log \mathcal{N}(T^n)\right)$$

$$\leq \liminf_{k \geq 1} \frac{1}{k} \log \mathcal{N}(S^k)$$

$$= \lim_{k \to \infty} \frac{1}{k} \log \mathcal{N}(S^k)$$

(since $\left(\frac{m \log \mathcal{N}(S^k)}{k}\right)_{k \in \mathbb{N}}$ is a sub-sequence of $\left(\frac{\log \mathcal{N}(S^n)}{n}\right)_{n \in \mathbb{N}}$). Moreover, we can estimate

$$m\left(\lim_{n \to \infty} \frac{1}{n} \log \mathcal{N}(T^n)\right)$$

$$\geq \lim_{n \to \infty} \frac{m}{n} \log \left(\mathcal{N}(S^{[\frac{n}{m}]})\mathcal{N}(T^{n - [\frac{n}{m}]m})\right) \quad \text{(by Lemma 4.5)}$$

$$= \lim_{n \to \infty} \frac{m}{n} \log \left(\mathcal{N}(S^{[\frac{n}{m}]}) + \min_{0 \leq i \leq m-1} \{\log \mathcal{N}(T^i)\}\right)$$

$$= \lim_{k \to \infty} \frac{1}{k} \log \mathcal{N}(S^k).$$

Comparing these last two inequalities we see that

$$\lim_{k \to \infty} \frac{1}{k} \log \mathcal{N}(S^k) = m\left(\lim_{n \to \infty} \frac{1}{n} \log \mathcal{N}(T^n)\right).$$

We shall now concentrate on the interval map $S : I \to I$. We denote by $\{J_r\}_{r=0}^M$ the intervals of monotonicity for $S : I \to I$ (with neighbouring intervals labelled consecutively).

Let $\alpha = \{U_r\}_{r=1}^m$ denote the open cover for I whose elements are open intervals formed from neighbouring intervals from $\{J_r\}$ (i.e. $U_1 = \mathrm{int}(J_1 \cup J_2)$, $U_2 = \mathrm{int}(J_1 \cup J_2 \cup J_3), \ldots, U_r = \mathrm{int}(J_{r-1} \cup J_r \cup J_{r+1}), \ldots, U_m = \mathrm{int}(J_{m-1} \cup J_m))$.

For $n \geq 1$, each non-degenerate interval $J_{i_0} \cap S^{-1} J_{i_1} \cap \ldots \cap S^{-(n-1)} J_{i_{(n-1)}}$ is now an interval of monotonicity for S^n and corresponds to at most 3^n elements of the cover

$$\vee_{r=0}^{n-1} S^{-r} \alpha$$
$$= \left\{U_{j_0} \cap S^{-1} U_{j_1} \cap \ldots \cap S^{-(n-1)} U_{j_{(n-1)}} : j_0, \ldots, j_{(n-1)} \in \{1, \ldots, N\}\right\}$$

(given by the at most three choices $\mathrm{int} J_{(i_r - 1)}$, $\mathrm{int} J_{i_r}$, and $\mathrm{int} J_{(i_r + 1)}$ contained in U_{i_r}, for $0 \leq r \leq n - 1$). In particular, we see that $\mathcal{N}(S^k) \leq$

$3^k N(\vee_{r=0}^{k-1} S^{-r}\alpha)$. By the definition of topological entropy in section 3.2 we have that $h(S) \geq h(S,\alpha)$ and then we may write

$$
\begin{aligned}
h(S) &\geq h(S,\alpha)\\
&= \lim_{k\to+\infty} \frac{1}{k} H\left(\vee_{i=0}^{k-1} S^{-i}\alpha\right)\\
&\geq \limsup_{k\to+\infty} \frac{1}{k} \log\left(\mathcal{N}(S^k)/3^n\right) \quad \text{(since } \mathcal{N}(S^k) \leq 3^k N(\vee_{r=0}^{k-1} S^{-r}\alpha))\\
&= \limsup_{k\to+\infty} \frac{1}{k} \log\mathcal{N}(S^k) - \log 3
\end{aligned}
$$

Finally, by Corollary 3.8.1 we know that $h(S) = mh(T)$ and so we have

$$
\begin{aligned}
h(T) &= \frac{h(S)}{m}\\
&\geq \frac{1}{m}\left(\limsup_{k\to+\infty} \frac{1}{k} \log\mathcal{N}(S^k)\right) - \frac{\log 3}{m}\\
&= \limsup_{n\to+\infty} \frac{1}{n} \log\mathcal{N}(T^n) - \frac{\log 3}{m}.
\end{aligned}
$$

Since $m \geq 1$ can be arbitrarily large, we get

$$
h(T) \geq \limsup_{n\to+\infty} \frac{1}{n} \log\mathcal{N}(T^n). \tag{4.3}
$$

Comparing (4.2) and (4.3) completes the proof. ∎

4.3 Markov maps

Consider a division of the interval $I = [0,1]$ into a finite number of closed sub-intervals $I_i = [x_i, x_{i+1}]$ $(i = 0,\ldots,k-1)$ with endpoints $0 = x_0 < x_1 < \ldots < x_k = 1$.

DEFINITION. We are interested in surjective maps $T : I \to I$ which are C^1 and monotone on each of the (open) intervals $\text{int}I_i = (x_{i-1}, x_i)$ and satisfy the following additional properties.

  (i) (*Piecewise expanding*) There exists $\beta > 1$ such that $|T'(x)| \geq \beta$, $\forall x \in I_i$ $(i = 1,\ldots,k)$;
  (ii) (*Markov property*) If $T(\text{int}I_i) \cap \text{int}I_j \neq \emptyset$ then $T(\text{int}I_i) \supset \text{int}I_j$ (for $i,j = 1,\ldots,k$).

(see figure 4.4)

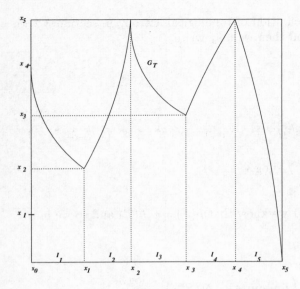

FIGURE 4.4. $\mathcal{G}_T$ for a Markov map

EXAMPLE. Consider the interval $I = [0, 1]$ and the map $T : I \to I$ defined by

$$T(x) = \begin{cases} 2x & \text{if } 0 \leq x \leq \frac{1}{2}, \\ 2x - 1 & \text{if } \frac{1}{2} < x \leq 1. \end{cases}$$

In this case we take $I_1 = [0, \frac{1}{2}]$ and $I_2 = [\frac{1}{2}, 1]$. It is easy to see that this map is piecewise expanding (where we can take $\beta = 2$ in (i)). Also, since $T(0, \frac{1}{2}) = T(\frac{1}{2}, 1) = (0, 1) \supset (0, \frac{1}{2}) \cup (\frac{1}{2}, 1)$ it is a simple matter to see that the Markov property holds.

TECHNICAL POINT FOR THE CAUTIOUS. These maps $T : I \to I$ may not be continuous at the points $x_1 < \ldots < x_{k-1}$, which sits a little uncomfortably with our earlier definitions (of topological entropy, etc.) for continuous maps. This can be rectified by the simple device of looking at the map T on the *disjoint* union $\cup_{i=1}^{k-1} I_i$ (i.e. treating the endpoints of different intervals as different points). To avoid this concern, we invite the reader to restrict his attention to continuous maps.

The following is a useful property for such maps to have.

DEFINITION. We say that $T : I \to I$ is *locally eventually onto* if for every interval $J \subset I$ there exists $n \geq 1$ with $T^n J = I$.

We want to associate to a piecewise expanding Markov interval map a $k \times k$ matrix A with entries either 0 or 1.

DEFINITION. We define a *transition matrix* A by

$$A(i,j) = \begin{cases} 1 & \text{if } T(\text{int} I_i) \supset \text{int} I_j, \\ 0 & \text{if } T(\text{int} I_i) \cap \text{int} I_j = \emptyset. \end{cases}$$

We can now define a *one-sided subshift of finite type* on the shift space of sequences,

$$X_A^+ = \left\{ x = (x_n)_{n \in \mathbb{Z}^+} \in \prod_{\mathbb{Z}^+} \{1, \ldots, k\} : A(x_n, x_{n+1}) = 1, \text{ for } n \geq 0 \right\},$$

by $\sigma : X_A^+ \to X_A^+$ with $\sigma((x_n)_{n \in \mathbb{Z}^+}) = (x_{n+1})_{n \in \mathbb{Z}^+}$ (i.e. all the terms in the sequence x are shifted one place to the left, except the zeroth term x_0 which is thrown away).

For $x = (x_n)_{n \in \mathbb{Z}^+}$ and $y = (y_n)_{n \in \mathbb{Z}^+}$ we define

$$N^+(x,y) = \min\{N \geq 0 : x_N \neq y_N\}.$$

A natural metric on this space is defined by $d(x,y) = \left(\frac{1}{2}\right)^{N^+(x,y)}$ for $x, y \in X_A^+$ ($x \neq y$).

LEMMA 4.7. *The space X_A^+ is compact and the map $\sigma : X_A^+ \to X_A^+$ is continuous. Moreover, every point $x \in X_A^+$ has at most k pre-images*

PROOF. The proofs of the first two assertions are almost identical to those for $\sigma : X_A \to X_A$ in chapter 1, and so we shall omit them.

For the last part, we observe that if $x = (x_n)_{n \in \mathbb{Z}^+} \in X_A^+$ and $y = (y_n)_{n \in \mathbb{Z}^+}$ satisfies $\sigma(y) = x$ then the values $y_1 = x_0, y_2 = x_1, \ldots, y_r = x_{r-1}, \ldots$ are all determined, and so y is completely specified up to the at most k possible choices $y_0 \in \{1, \ldots, k\}$ (such that $A(y_0, x_0) = 1$).

∎

The next proposition shows that $\sigma : X_A^+ \to X_A^+$ gives a lot of information about $T : I \to I$.

PROPOSITION 4.8. *There is continuous map $\pi : X_A^+ \to I$ such that*
  (i) *π is surjective and $\pi \circ \sigma = T \circ \pi$ (i.e. π is a semi-conjugacy),*
 (ii) *points $z \in I$ have exactly one or two pre-images in X_A^+ (i.e. $\forall z \in I$ the set $E(z) = \{x \in X_A^+ : \pi(x) = z\}$ consists of either one or two points),*
(iii) *the set of points $z \in I$ such that $E(z)$ consists of more than one point is contained in the countable set $\cup_{n \in \mathbb{Z}^+} T^{-n}\{x_0, \ldots, x_k\}$,*
(iv) *if $T : I \to I$ is locally eventually onto then A is aperiodic (i.e. $\exists n \geq 1$, $\forall 1 \leq i, j \leq k$, $A^n(i,j) > 0$).*

PROOF. We want to define the map $\pi : X_A^+ \to I$ by

$$\pi(w) = \cap_{n=0}^{\infty} \text{cl}\left(T^{-n}\text{int}(I_{w_n})\right)$$

where $w = (w_n)_{n \in \mathbb{Z}^+} \in X_A^+$ (i.e. $\pi(w)$ should correspond to a point $x \in I$ such that $T^n x \in I_{w_n}$ for $n \geq 0$). The use of cl (closure) and int (interior) helps to avoid tiresome problems with the endpoints of intervals.

We first show that this map is well-defined. For any $N \geq 0$ we have that $J_N(w) = \cap_{n=0}^{N} \text{cl}\left(T^{-n}\text{int}(I_{w_n})\right)$ is non-empty and closed. Since $J_0(w) \supset J_1(w) \supset \ldots \supset J_k(w) \supset \ldots$ is a nested sequence of closed sets we have by compactness that $\cap_{k \in \mathbb{Z}^+} J_k(w) \neq \emptyset$. Moreover, since $T : I \to I$ is piecewise expanding we see that

$$\text{diam}(J_n(x)) \leq \frac{1}{\beta}\text{diam}(J_{n-1}(x)) \leq \ldots$$

$$\ldots \leq \frac{1}{\beta^{n-1}}\text{diam}(J_1(x)) \leq \frac{1}{\beta^n} \to 0$$

as $n \to +\infty$. In particular, this intersection consists of a single point, which we take to be $\pi(w)$.

To see that $\pi : X_A^+ \to I$ is continuous, we begin by choosing for each $\epsilon > 0$ an integer $n \geq 1$ such that $\frac{1}{\beta^n} < \epsilon$. If $d(x,y) \leq \frac{1}{2^n}$ then $x_i = y_i$ for $i = 0, \ldots, n-1$ and $\pi(x), \pi(y) \in J_n(x)$ (by definition of $J_n(x)$). In particular, $|\pi(x) - \pi(y)| \leq \text{diam} J_n(x) \leq \frac{1}{\beta^n} < \epsilon$.

(i) To see that π is surjective, we need only observe that for every point x in the dense set $\cup_{n \in \mathbb{Z}^+} T^{-n}\{x_0, \ldots, x_k\}$ we have a unique choice $(w_n)_{n \in \mathbb{Z}^+} \in X_A^+$ with $T^n(x) \in \text{int}(I_{w_n})$ for $n \geq 0$, and thus $\pi\left((w_n)_{n \in \mathbb{Z}^+}\right) = x$. Since the image of π is compact and contains a dense set, we conclude that π is surjective.

To see that $\pi \circ \sigma = T \circ \pi$ observe that

$$\pi \circ \sigma(w_n)_{n \in \mathbb{Z}^+} = \pi\left((w_{n+1})_{n \in \mathbb{Z}^+}\right)$$
$$= \cap_{n \in \mathbb{Z}^+} \text{cl}\left(T^{-n}\text{int}(I_{w_{n+1}})\right)$$
$$= (T \circ \pi)\left((w_n)_{n \in \mathbb{Z}^+}\right).$$

(ii) Assume that $z = \pi(w) = \pi(y)$ with $w = (w_n)_{n \in \mathbb{Z}^+} \neq y = (y_n)_{n \in \mathbb{Z}^+}$. If $w_0 \neq y_0$ then $z \in I_{w_0} \cap I_{y_0} \neq \emptyset$, i.e. z must be one of the endpoints $x_0, \ldots, x_n$. However, if $z = x_i$, say, then the choice of either $w_0 = i$ or $w_0 = i + 1$ uniquely determines the rest of the sequence $w = (w_n)_{n \in \mathbb{Z}^+}$ with $\pi(w) = x$. Thus $E(z)$ can have cardinality at most two.

More generally, if $w_i = y_i$ for $i = 0, \ldots, r-1$ but $w_r \neq y_r$ then $T^r z \in I_{w_r} \cap I_{y_r}$. Thus, $T^r z = x_j$, say. Moreover, the choice of $w_r = j$ or $w_r = j+1$ uniquely determines the rest of the sequence $w = (w_n)_{n \in \mathbb{Z}^+}$ with $\pi(w) = x$.

(iii) From the proof of (ii) we see that we require that $z \in \cup_{n=0}^{\infty} T^{-n}\{x_0, \ldots, x_k\}$. Since T is Markov this is a countable set.

(iv) For each $1 \leq i \leq k$ we can use the definition of locally eventually onto to choose $n_i \geq 1$ with $T^{n_i} I_i = I$. In particular, for $n \geq \max\{n_i : 1 \leq i \leq k\}$ we see that for any $1 \leq i \leq k$ we have $T^n I_i = T^{n-n_i}(T^{n_i} I_i) = T^{n-n_i}(I) = I$. In particular, this means that $\forall 1 \leq i, j \leq k$ we can find $x \in I_i \cap T^{-n} I_j$. If we choose $i_1, \ldots, i_{n-1} \in \{1, \ldots, k\}$ such that $T^r x \in I_{i_r}$ $(r = 1, \ldots, n-1)$ then we see that

$$A^n(i, j) \geq A(i, i_1) A(i_1, i_2) \ldots A(i_{n-1}, j) = 1.$$

This shows that A is aperiodic.

∎

The usefulness of Proposition 4.8 lies in converting relatively easy results for $\sigma : X_A^+ \to X_A^+$ into corresponding results for $T : I \to I$.

THEOREM 4.9. *If $T : I \to I$ is locally eventually onto then $h(T) = \log \lambda_1$ where λ_1 is the unique maximal eigenvalue for the matrix A.*

PROOF OF THEOREM 4.9. Since $\pi : X_A^+ \to I$ is a surjective semi-conjugacy between $\sigma : X_A^+ \to X_A^+$ and $T : I \to I$ we have by Proposition 3.5 that $h(T) \leq h(\sigma)$. To get the reverse inequality we need only argue as in the proof of Theorem 4.6. This shows that $h(T) = h(\sigma)$.

To complete the proof, we only need the following sublemma.

SUBLEMMA 4.9.1. $h(\sigma) = \log \lambda_1$.

PROOF OF SUBLEMMA 4.9.1. The proof is identical to that of Proposition 3.8.

∎

An algebraic number is one which is the root of a polynomial with integer entries. There are only countably many algebraic numbers.

COROLLARY 4.9.1. *The value $e^{h(T)}$ is an algebraic number.*

PROOF. Observe that $\lambda_1 = e^{h(\sigma)}$ is an eigenvalue for A and so, in particular, a root of the characteristic polynomial $p(z) = \det(zI - A)$ which has integer coefficients (since the matrix A has integer entries). By Theorem 4.9 we have $e^{h(\sigma)} = e^{h(T)}$.

∎

DEFINITION. For each $n \geq 1$ we let $\text{Fix}(T^n) = \{x \in I : T^n x = x\}$ denote the set of fixed point for T^n and let $\text{Card}(\text{Fix}(T^n))$ be the cardinality of this set.

THEOREM 4.10. $\lim_{n \to +\infty} \frac{\text{Card}(\text{Fix}(T^n))}{e^{nh(T)}} = 1.$

PROOF OF THEOREM 4.10. We first establish the analogous result in the easier context of $\sigma : X_A^+ \to X_A^+$. By Sublemma 4.9.1. the topological entropy of this map is $\log \lambda_1$.

SUBLEMMA 4.10.1. $\lim_{n \to +\infty} \frac{\text{Card} Fix(\sigma^n)}{\lambda_1^n} = 1$

PROOF OF SUBLEMMA 4.10.1. A fixed point $x = (x_k)_{k \in \mathbb{Z}^+}$ for $\sigma^n : X_A^+ \to X_A^+$ is a sequence for which $x_k = x_{k+n} = x_{k+2n} = \ldots$, for $k = 0, 1, \ldots, n-1$. Thus $\text{Card}(\text{Fix}(\sigma^n))$ is exactly the number of strings $(x_0, x_1, \ldots, x_{n-1})$ with $A(x_0, x_1) = A(x_1, x_2) = \ldots = A(x_{n-2}, x_{n-1}) = A(x_{n-1}, x_0) = 1$. However, this is given by $\text{trace}(A^n) = \lambda_1^n + \lambda_2^n + \ldots + \lambda_k^n$, where λ_i $(i = 1, \ldots, k)$ are the eigenvalues for A.

Since by the Perron-Frobenius theorem we have $|\lambda_i| < \lambda_0$ for $i = 2, \ldots, k$ we see that

$$\lim_{n \to +\infty} \frac{\text{Card}(\text{Fix}(\sigma^n))}{\lambda_1^n} = \lim_{n \to +\infty} \frac{\lambda_1^n + \ldots + \lambda_k^n}{\lambda_1^n} = 1. \qquad (4.4)$$

∎

By Proposition 4.8 (iii) there is a bijection between $\text{Fix}(\sigma^n)$ and $\text{Fix}(T^n)$, with the possible exception of the finite set $\text{Fix}(T^n) \cap \{x_1, \ldots, x_k\}$. Thus $|\text{Fix}(\sigma^n) - \text{Fix}(T^n)| \leq k$ and so by (4.4)

$$1 = \lim_{n \to +\infty} \frac{\text{Card}(\text{Fix}(\sigma^n)) - k}{e^{nh(\sigma)}} \leq \liminf_{n \to +\infty} \frac{\text{Card}(\text{Fix}(T^n))}{e^{nh(T)}}$$

$$\leq \limsup_{n \to +\infty} \frac{\text{Card}(\text{Fix}(T^n))}{e^{nh(T)}} \leq \lim_{n \to +\infty} \frac{\text{Card}(\text{Fix}(\sigma^n)) + k}{e^{nh(\sigma)}} = 1.$$

This completes the proof.

∎

4.4 Comments and references

Sharkovski's theorem on periodic points can be found in [8], [9], [2] or [5] (for period 3 orbits).

The interpretation of the topological entropy as the growth rate of monotone intervals is due to Milnor and Thurston [6].

The use of subshifts of finite type to study Markov interval maps is a simple analogue of Markov Partitions and symbolic dynamics for Axiom A diffeomorphisms (cf. references to chapter 5). If we drop the Markov assumption then there is a construction due to Hofbauer [4] of Markov extensions.

Other important aspects of interval maps which we have omitted are dealt with in [2] (kneading theory), [1] (renormalization) and [3] (period doubling).

References

1. P. Collet and J.-P. Eckmann, *Iterated Maps of the Interval as Dynamical Systems*, Birkhäuser, Boston Mass., 1980.
2. R. Devaney, *An Introduction to Chaotic Dynamical Systems*, Addison-Wesley, New York, 1989.
3. W. de Melo and S. van Strien, *One Dimensional Dynamics*, Springer, Berlin, 1993.
4. F. Hofbauer, *On the intrinsic ergodicity of piecewise monotonic transformations*, Israel J. Math. **34** (1979), 213-237.
5. T. Li and J. Yorke, *Period three implies chaos*, Amer. Math. Monthly **82** (1975), 985-992.
6. J. Milnor and W. Thurston, *On iterated maps of the interval*, Dynamical Systems (J.Alexander, ed.), Lecture Notes in Mathematics 1342, Springer, Berlin, 1988, pp. 465-563.
7. F. Schweiger, *Ergodic Theory of Fibred Markov Systems and Metric Number Theory*, O.U.P., Oxford, 1995.
8. Sharkovski, *Coexistence of cycles of a continuous map of the line into itself*, Ukrain. Mat. Zh. **16** (1964), 61-71.
9. P. Stefan, *A theorem of Sharkovski on the existence of periodic orbits of continuous endomorphisms on the real line*, Commun. Math. Phys. **54** (1977), 237-248.

CHAPTER 5

HYPERBOLIC TORAL AUTOMORPHISMS

We want to consider a simple class of homeomorphisms whose dynamical behaviour illustrates many of the ideas we have discussed.

5.1 Definitions

Let $SL(2, \mathbb{Z})$ be the set of all 2×2 matrices $A = \begin{pmatrix} a & b \\ c & d \end{pmatrix}$, where $a, b, c, d \in \mathbb{Z}$ and $\mathrm{Det}(A) = ad - bc = \pm 1$. Each such matrix A gives a linear map on $\mathbb{R}^2$ by $\begin{pmatrix} x_1 \\ x_2 \end{pmatrix} \mapsto A \begin{pmatrix} x_1 \\ x_2 \end{pmatrix}$.

Let $\mathbb{T}^2 = \mathbb{R}^2/\mathbb{Z}^2$ be the two-dimensional torus and we define a *linear toral automorphism* $T : \mathbb{T}^2 \to \mathbb{T}^2$ by $T(x_1, x_2) = (ax_1 + bx_2, cx_1 + dx_2) \pmod 1$ (see figure 5.1). We say that $T : \mathbb{T}^2/\mathbb{Z}^2 \to \mathbb{T}^2/\mathbb{Z}^2$ is *hyperbolic* if A does not have eigenvalues of modulus 1.

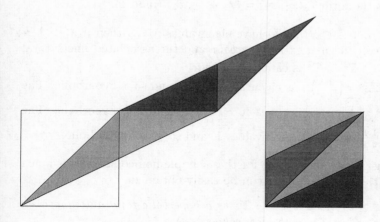

FIGURE 5.1. The map $A : \mathbb{R}^2 \to \mathbb{R}^2$ projects down to a map $T : \mathbb{T}^2 \to \mathbb{T}^2$

PROPOSITION 5.1. *A hyperbolic linear toral automorphism* $T : \mathbb{T}^2 \to \mathbb{T}^2$ *is a homeomorphism*

PROOF. The map T is clearly continuous since if $|x_1 - y_1|, |x_2 - y_2| < \epsilon$ then $|(T(x_1, x_2))_1 - (T(y_1, y_2))_1| \leq (|a| + |b|)\epsilon$ and $|(T(x_1, x_2))_2 - (T(y_1, y_2))_2| \leq (|c| + |d|)\epsilon$.

To see that T is invertible we note that if we write the inverse matrix

$$A^{-1} = \begin{pmatrix} \frac{d}{ad-bc} & \frac{-b}{ad-bc} \\ \frac{-c}{ad-bc} & \frac{a}{ad-bc} \end{pmatrix}$$

then since $ad - bc = \pm 1$ we see that $A^{-1} \in SL(2, \mathbb{Z})$. The inverse to T : $\mathbb{T}^2 \to \mathbb{T}^2$ is then the linear toral automorphism associated to A^{-1}, i.e.

$$T^{-1}(x_1, x_2) = (\frac{d}{ad - bc}x_1 + \frac{-b}{ad - bc}x_2, \frac{-c}{ad - bc}x_1 + \frac{a}{ad - bc}x_2) \text{ (mod 1)},$$

∎

We give some examples.

EXAMPLES.
  (i) Let $A_1 = \begin{pmatrix} 2 & 1 \\ 1 & 1 \end{pmatrix}$; then $\text{Det}(A_1) = 1$ and the associated map T_1 : $\mathbb{T}^2 \to \mathbb{T}^2$ takes the form $T_1(x_1, x_2) = (2x_1 + x_2, x_1 + x_2)$ (mod 1).
  (ii) Let $A_2 = \begin{pmatrix} 57 & 2 \\ 85 & 3 \end{pmatrix}$; then $\text{Det}(A_2) = 1$ and the associated map T_2 : $\mathbb{T}^2 \to \mathbb{T}^2$ takes the form $T_2(x_1, x_2) = (57x_1 + 2x_2, 85x_1 + 3x_2)$ (mod 1).
  (iii) Let $A_3 = \begin{pmatrix} 1 & 1 \\ 0 & 1 \end{pmatrix}$ then $\text{Det}(A_3) = 1$ and the associated map $T_3 : \mathbb{T}^2 \to \mathbb{T}^2$ takes the form $T_3(x_1, x_2) = (x_1 + x_2, x_2)$ (mod 1).

DEFINITION. Let $A \in SL(2, \mathbb{Z})$ have eigenvalues λ_1, λ_2 then if $\lambda_1 > 1 > \lambda_2 (= \pm\frac{1}{\lambda_1})$ we call the matrix A *hyperbolic* and the associated linear toral transformation $T : \mathbb{T}^2 \to \mathbb{T}^2$ is called *hyperbolic*.

The matrix $A_1 = \begin{pmatrix} 2 & 1 \\ 1 & 1 \end{pmatrix}$ has eigenvalues $\frac{3 \pm \sqrt{5}}{2}$, and so is hyperbolic. The matrix $A_2 = \begin{pmatrix} 57 & 2 \\ 85 & 3 \end{pmatrix}$ has eigenvalues $30 \pm \frac{1}{2}\sqrt{3596}$, and so is hyperbolic. The matrix $A_3 = \begin{pmatrix} 1 & 1 \\ 0 & 1 \end{pmatrix}$ has both eigenvalues 1, and so is *not* hyperbolic.

The following results shows that for these simple homeomorphisms much information on the periodic points can be easily obtained.

PROPOSITION 5.2. *Let $T : \mathbb{T}^2 \to \mathbb{T}^2$ be a hyperbolic toral automorphism associated to $A \in SL(2, \mathbb{Z})$ (with eigenvalues λ_1, λ_2).*

  (i) *The periodic points have rational coordinates in $\mathbb{T}^2$ (i.e. they are of the form $\left(\frac{p_1}{q}, \frac{p_2}{q}\right)$ with natural numbers $0 \le p_1, p_2 < q$).*
  (ii) *We have $\text{Card}\{x \in \mathbb{T}^2 : T^n x = x\} = |\lambda_1^n + \lambda_2^n - 2|$.*

PROOF. (i) If $(x_1, x_2) = \left(\frac{p_1}{q}, \frac{p_2}{q}\right)$ is a point with rational co-ordinates then we can write $T^n(x_1, x_2) + \mathbb{Z}^2 = \left(\frac{p_1^{(n)}}{q}, \frac{p_2^{(n)}}{q}\right) + \mathbb{Z}^2$, where the integers

$0 \leq p_1^{(n)}, p_2^{(n)} \leq q - 1$ are given by $\begin{pmatrix} p_1^{(n)} \\ p_2^{(n)} \end{pmatrix} = A^n \begin{pmatrix} p_1 \\ p_2 \end{pmatrix}$. Since there are at most q^2 distinct choices $p_1^{(n)}, p_2^{(n)}$ we can choose $q^2 + 1 \geq n_1 > n_0 > 0$ with $p_1^{(n_0)} = p_1^{(n_1)}$ and $p_2^{(n_0)} = p_2^{(n_1)}$. In particular, this means that $A^{n_1} \begin{pmatrix} p_1 \\ p_2 \end{pmatrix} = A^{n_0} \begin{pmatrix} p_1 \\ p_2 \end{pmatrix}$. This corresponds to the identity $T^{n_1}(x_1, x_2) = T^{n_0}(x_1, x_2)$ and so we conclude that $T^{n_1 - n_0}(x_1, x_2) = (x_1, x_2)$, i.e. (x_1, x_2) is periodic.

Conversely, assume that $T^n(x_1, x_2) = (x_1, x_2)$ is a periodic point; then

$$A^n \begin{pmatrix} x_1 \\ x_2 \end{pmatrix} = \begin{pmatrix} x_1 \\ x_2 \end{pmatrix} + \begin{pmatrix} n_1 \\ n_2 \end{pmatrix}, \tag{5.1}$$

for some $n_1, n_2 \in \mathbb{Z}$. Since A^n does not have 1 as an eigenvalue, the matrix $A^n - I$ is invertible and solutions to (5.1) are of the form $\begin{pmatrix} x_1 \\ x_2 \end{pmatrix} = (A^n - I)^{-1} \begin{pmatrix} n_1 \\ n_2 \end{pmatrix}$. Since $A^n - I$ has entries in $\mathbb{Z}$ the matrix $(A^n - I)^{-1}$ has entries in $\mathbb{Q}$, and we can conclude the same for the values x_1, x_2.

This completes the proof of part (i).

(ii) If we write $A^n = \begin{pmatrix} a_n & b_n \\ c_n & d_n \end{pmatrix}$ then we can define a map $S = (A^n - I)$: $\mathbb{R}^2 \to \mathbb{R}^2$ by $S : (x_1, x_2) \to ((a_n - 1)x_1 + b_n x_2, c_n x_1 + (d_n - 1)x_2)$. This maps the unit square $[0, 1) \times [0, 1)$ onto the rhombus $\mathcal{R} = \{\alpha u + \beta v : 0 \leq \alpha, \beta < 1\}$ where $u = S(0, 1)$ and $v = S(1, 0)$ (see figure 5.2).

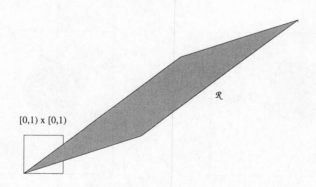

$[0,1) \times [0,1)$

$\mathcal{R}$

FIGURE 5.2.   The number of points in $\mathbb{Z}^2 \cap \mathcal{R}$ is $\mathrm{Card}(\mathrm{Fix}(T^n))$

A fixed point $T^n(x_1, x_2) = (x_1, x_2)$ corresponds to a solution to the identity (5.1) or, equivalently, to solutions $S(x_1, x_2) \in \mathbb{Z}^2 \cap \mathcal{R}$ (where $(x_1, x_2) \in [0, 1] \times [0, 1]$). Due to the shape of a rhombus the number of such solutions is precisely the area of $\mathcal{R}$, which is $|\mathrm{Det}(A^n - I)| = |\lambda_1^n + \lambda_2^n - 2|$. This completes the proof of part (ii).

∎

5.2 Entropy for hyperbolic toral automorphisms

The main result on the entropy is the following.

THEOREM 5.3. *The entropy of the hyperbolic toral automorphism is given by $h(T) = \log \lambda_1$.*

PROOF. We fix $\epsilon > 0$. We can cover $\mathbb{T}^2$ by ϵ-balls centred at a finite set of points $\{(x_1^1, x_2^1), \dots, (x_1^k, x_2^k)\}$,

$$B\left((x_1^i, x_2^i), \epsilon\right) = \{(z_1, z_2) \in \mathbb{T}^2 : |(z_1, z_2) - (x_1, x_2)| < \epsilon\}$$

(cf. Figure 5.3.) In particular, we can arrange $k \leq \frac{4}{\epsilon^2}$.

Around each point $x^i = (x_1^i, x_2^i), i = 1, \dots, k$, we can describe a "box"

$$\mathrm{Box}(x_1^i, x_2^i) = \{(x_1^i, x_2^i) + \alpha v_1 + \beta v_2 : -\epsilon \leq \alpha, \beta \leq \epsilon\}$$

where v_1 and v_2 are the eigenvectors for A (with $|v_1| = |v_2| = 1$) corresponding to $|\lambda_1| > 1$ and $|\lambda_2| < 1$, respectively. For each $n \geq 1$ and $1 \leq i \leq k$ we can consider the finite subset of $\mathrm{Box}(x^i)$ consisting of the points

$$R(x_1^i, x_2^i) = \{(x_1^i, x_2^i) + \frac{j\epsilon}{|\lambda_1|^n} v_1 : j = -[|\lambda_1|^n], \dots, [|\lambda_1|^n]\}.$$

This set has cardinality $2[|\lambda_1|^n] + 1$.

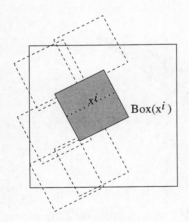
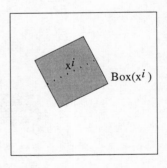

FIGURE 5.3. Construction an $(n, 2\epsilon)$-spanning set and an (n, ϵ)-separating set

SUBLEMMA 5.3.1. $R = \cup_{i=1}^k R(x^i)$ *is an $(n, 2\epsilon)$-spanning set (see figure 5.3).*

PROOF. For any point $(z_1, z_2) \in \mathbb{R}^2/\mathbb{Z}^2$ we can choose $|(x_1^i, x_2^i) - (z_1, z_2)| < \epsilon$, for some $i = 1, \dots, k$. In particular, this implies that $(z_1, z_2) \in \mathrm{Box}(x_1^i, x_2^i)$ and so we can write $(z_1, z_2) = (x_1^i, x_2^i) + \alpha v_1 + \beta v_2$ for some $-\epsilon \leq \alpha, \beta \leq \epsilon$.

If we choose $-[|\lambda_1|^n] \le j \le [|\lambda_1|^n]$ with $|\alpha - \frac{j\epsilon}{|\lambda_1|^n}| \le \frac{\epsilon}{2|\lambda_1|^n}$ then we can select $(w_1, w_2) = (x_1^i, x_2^i) + \frac{j\epsilon}{|\lambda_2|^n} v_1 \in R(x_1^i, x_2^i)$.

For any $0 \le r \le n$ we can write

$$T^r(z_1, z_2) = T^r(w_1, w_2) + \left(\alpha - \frac{j\epsilon}{|\lambda_1|^n}\right) A^r v_1 + \beta A^r v_2$$

$$= T^r(w_1, w_2) + \left(\alpha - \frac{j\epsilon}{|\lambda_1|^n}\right) \lambda_1^r v_1 + \beta \lambda_2^r v_2$$

and, in particular,

$$|T^r(z_1, z_2) - T^r(w_1, w_2)| \le |\left(\alpha - \frac{j\epsilon}{|\lambda_1|^n}\right)| \cdot |\lambda_1|^r \cdot |v_1| + \beta \lambda_2^r |v_2|$$

$$\le \frac{\epsilon|\lambda_1|^r}{2|\lambda_1|^n} + \epsilon \le \frac{\epsilon}{2} + \epsilon = 2\epsilon.$$

This completes the proof of Sublemma 5.3.1. ∎

The cardinality of the $(n, 2\epsilon)$-spanning set R is at most $k(2[|\lambda_1|^n] + 1)$ we see that this is an upper bound on the least cardinality $r(n, \epsilon)$ of $(n, 2\epsilon)$-spanning sets. By Proposition 3.11 we can write

$$h(T) = \lim_{\epsilon \to 0} \lim_{n \to +\infty} \frac{1}{n} \log r(n, \epsilon)$$

$$\le \lim_{\epsilon \to 0} \lim_{n \to +\infty} \frac{1}{n} \log \left(\frac{16(2[|\lambda_1|^n] + 1)}{\epsilon^2}\right) \tag{5.2}$$

$$= \lim_{\epsilon \to 0} \log |\lambda_1| = \log |\lambda_1|.$$

To get the reverse inequality, we fix a point $(x_1, x_2) \in \mathbb{T}^2$. For $\epsilon > 0$ and $n \ge 1$ we can consider the subset

$$S = \{(x_1, x_2) + \frac{j2\epsilon}{|\lambda_1|^n} v_1 : j = -[|\lambda_1|^n], \dots, [|\lambda_1|^n]\}.$$

SUBLEMMA 5.3.2. *S is an (n, ϵ)-separated set (see Figure 5.3).*

PROOF. Any two distinct points in S will be of the form $(u_1, u_2) = (x_1, x_2) + \frac{i2\epsilon}{|\lambda_1|^n} v_1$, $(w_1, w_2) = (x_1, x_2) + \frac{j2\epsilon}{|\lambda_1|^n} v_1$, with $-[|\lambda_1|^n] \le i, j \le [|\lambda_1|^n]$. For $0 \le r \le n - 1$ we have

$$|T^r(u_1, u_2) - T^r(w_1, w_2)| = |\frac{(j-i)2\epsilon}{|\lambda_1|^n} T^r(v_1)|$$

$$= 2|j - i| \frac{\epsilon}{|\lambda_1|^{n-r}}.$$

In particular, for some $0 \leq r \leq n-1$ we have that $|T^n(u_1, u_2) - T^n(w_1, w_2)| > \epsilon$. This completes the proof of Sublemma 5.3.2.

∎

The cardinality of the (n, ϵ)-separated set S is $2[|\lambda_1|^n] + 1$ and so this gives a lower bound on the maximal cardinality $s(n, \epsilon)$ of (n, ϵ)-separated sets. By Proposition 3.11 we have that

$$
\begin{aligned}
h(T) &= \lim_{\epsilon \to 0} \lim_{n \to +\infty} \frac{1}{n} \log s(n, \epsilon) \\
&\geq \lim_{\epsilon \to 0} \lim_{n \to +\infty} \frac{1}{n} \log \left(2[|\lambda_1|^n] + 1 \right) \\
&= \lim_{\epsilon \to 0} \left(\log |\lambda_1| \right) = \log |\lambda_1|.
\end{aligned}
\tag{5.3}
$$

The two complementary inequalities (5.2) and (5.3) combine to show that $h(T) = \log |\lambda_1|$. This completes the proof of Theorem 5.3.

∎

COROLLARY 5.3.1.

$$
\lim_{n \to +\infty} \frac{1}{e^{nh(T)}} \, Card\left(Fix(T^n)\right) = 1.
$$

PROOF. This is immediate from Proposition 5.2 (ii) and Theorem 5.3

∎

5.3 Shadowing and semi-conjugacy

We begin with a very useful property for hyperbolic toral automorphisms.

DEFINITION. Let $\delta > 0$. We call a sequence of points $(x_n)_{n \in \mathbb{Z}} \in \mathbb{T}^2$ a δ-pseudo-orbit if $|T(x_n) - x_{n+1}| < \delta$, for all $n \in \mathbb{Z}$.

THEOREM 5.4 (SHADOWING PROPERTY). For all $\epsilon > 0$ there exists $\delta > 0$ such that any δ-pseudo-orbit $(x_n)_{n \in \mathbb{Z}}$ is ϵ-close to a true orbit $(T^n(x))_{n=-\infty}^{\infty}$ (in the sense that $\forall n \in \mathbb{Z}$, $|T^n(x) - x_n| < \epsilon$).

Furthermore, if $\delta > 0$ is sufficiently small then there is a unique point x with this property.

PROOF. Given $\epsilon > 0$ we choose $\delta > 0$ sufficiently small that any vector $v \in \mathbb{R}^2$ with $|v| \leq \delta$ can be written in the form $v = \alpha v_1 + \beta v_2$ with $|\alpha|, |\beta| < \frac{\epsilon}{3}(1 - \frac{1}{|\lambda_1|})$.

Let $(x_n)_{n \in \mathbb{Z}}$ be any δ-pseudo-orbit. By our choice of δ, for each $n \in \mathbb{Z}$ we can write $T(x_{n-1}) = x_n - \alpha_n v_1 - \beta_n v_2$, where $|\alpha_n|, |\beta_n| \leq \frac{\epsilon}{3}(1 - \frac{1}{|\lambda_1|})$. Iterating this identity gives that $T^2(x_{n-2}) = x_n - (\alpha_n + \alpha_{n-1}\lambda_1) v_1 - (\beta_n + \beta_{n-1}\lambda_2) v_2$ and proceeding inductively we get that

$$
x_n = \begin{cases} T^n(x_0) + \left(\sum_{i=1}^{n} \alpha_i \lambda_1^{n-i} \right) v_1 + \left(\sum_{i=1}^{n} \beta_i \lambda_2^{n-i} \right) v_2 & \text{for } n \geq 1, \\ T^n(x_0) + \left(\sum_{i=-n}^{-1} \alpha_i \lambda_1^{n-i} \right) v_1 + \left(\sum_{i=-n}^{-1} \beta_i \lambda_2^{n-i} \right) v_2 & \text{for } n \leq -1. \end{cases}
$$

We can now define $x = x_0 + \left(\sum_{i=1}^{\infty} \alpha_i \lambda_1^{-i} \right) v_1 + \left(\sum_{i=-\infty}^{-1} \beta_i \lambda_2^{-i} \right) v_2$. It is easy to see that $|x - x_0| \leq (\sup_i |\alpha_i| + \sup_i |\beta_i|)/(1 - \frac{1}{|\lambda_1|}) \leq \frac{2}{3}\epsilon$. We observe that $T^n(x) = T^n(x_0) + \lambda_1^n \left(\sum_{i=1}^{\infty} \lambda_1^{-i} \alpha_i \right) v_1 + \lambda_2^n \left(\sum_{i=-\infty}^{-1} \lambda_2^{-i} \beta_i \right) v_2$, for any $n \in \mathbb{Z}$.

For $n \geq 1$ we have that

$$
\begin{aligned}
T^n(x) - x_n = &\left(\lambda_1^n \sum_{i=1}^{\infty} \alpha_i \lambda_1^{-i} - \sum_{i=1}^{n} \alpha_i \lambda_1^{n-i} \right) v_1 \\
&+ \left(\lambda_2^n \sum_{i=-\infty}^{-1} \beta_i \lambda_2^{-i} - \sum_{i=1}^{n} \beta_i \lambda_2^{n-i} \right) v_2.
\end{aligned}
\tag{5.4}
$$

The coefficient of v_1 in (5.4) is $\sum_{i=n+1}^{\infty} \alpha_i \lambda^{n-i}$, which is bounded by $(\sup_i |\alpha_i|)/(1 - \frac{1}{|\lambda_1|}) \leq \epsilon/3$. The coefficient of v_2 in (5.4) can be crudely bounded by $2(\sup_i |\beta_i|)/(1 - \frac{1}{|\lambda_1|}) \leq 2\epsilon/3$. Combining these estimates shows that $|T^n(x) - x_n| \leq \frac{2\epsilon}{3} + \frac{\epsilon}{3} = \epsilon$, for all $n \geq 1$.

For $n \leq -1$ we have that

$$
\begin{aligned}
T^n(x) - x_n = &\left(\lambda_1^n \sum_{i=1}^{\infty} \alpha_i \lambda_1^{-i} - \sum_{i=n}^{-1} \alpha_i \lambda_1^{n-i} \right) v_1 \\
&+ \left(\lambda_2^n \sum_{i=-\infty}^{-1} \beta_i \lambda_2^{-i} - \sum_{i=n}^{-1} \beta_i \lambda_2^{n-i} \right) v_2.
\end{aligned}
\tag{5.5}
$$

The coefficient of v_2 is $\sum_{i=-\infty}^{n+1} \beta_i \lambda_2^{n-i}$, which is bounded by $(\sup_i |\beta_i|) \frac{1}{1 - \frac{1}{|\lambda_1|}} \leq \frac{\epsilon}{3}$, and the coefficient of v_1 can be crudely dominated by $(\sup_i |\alpha_i|) \frac{2}{1 - \frac{1}{|\lambda_1|}} \leq 2\epsilon/3$. Thus $|T^n(x) - x_n| \leq \frac{\epsilon}{3} + \frac{2\epsilon}{3} = \epsilon$, for all $n \leq -1$.

This shows that the pseudo-orbit $(x_n)_{n \in \mathbb{Z}}$ is ϵ-close to the orbit $(T^n x)_{n \in \mathbb{Z}}$.

Consider two points $x, y \in \mathbb{T}^2$ which are ϵ-close to the same pseudo-orbit $(x_n)_{n \in \mathbb{Z}}$. To complete the proof of the theorem we need to show that providing δ is sufficiently small this property implies $x = y$.

By the triangle inequality we have that $|T^n(y) - T^n(x)| \leq |T^n(y) - x_n| + |x_n - T^n(x)| \leq 2\epsilon$.

Provided the separation $|x - y| < 2\epsilon$ is small we can write $y = x + \alpha v_1 + \beta v_2$ and then $T^n(y) = T^n(x) + \alpha \lambda_1^n v_1 + \beta \lambda_2^n v_2$, for any $n \in \mathbb{Z}$. If $\alpha \neq 0$ then $|T^n(y) - T^n(x)| \geq |\alpha||\lambda_1|^n - |\beta||\lambda_2|^n > 2\epsilon$, for some $n \geq 1$, giving a contradiction. If $\beta \neq 0$ then $|T^{-n}(y) - T^{-n}(x)| \geq |\beta||\lambda_1|^n - |\alpha||\lambda_2|^n > 2\epsilon$, for some $n \geq 1$, giving a contradiction.

This completes the proof. ∎

Let $T : \mathbb{T}^2 \to \mathbb{T}^2$ be a hyperbolic toral automorphism and let $S : \mathbb{T}^2 \to \mathbb{T}^2$ be a second homeomorphism. We can write these maps in terms of their co-ordinates:

$$\begin{cases} T(x_1, x_2) = (T_1(x_1, x_2), T_2(x_1, x_2)) \text{ and} \\ S(x_1, x_2) = (S_1(x_1, x_2), S_2(x_1, x_2)). \end{cases}$$

DEFINITION. Given $\delta > 0$ we say that S and T are *uniformly* δ-close if

$$\sup_{(x_1, x_2) \in \mathbb{T}^2} \{|T_1(x_1, x_2) - S_1(x_1, x_2)|, |T_2(x_1, x_2) - S_2(x_1, x_2)|\} < \delta.$$

The next result tells that any homeomorphism close to a hyperbolic toral automorphism $T : \mathbb{T}^2 \to \mathbb{T}^2$ (in the above sense) must be semi-conjugate to T.

THEOREM 5.5. *Let $T : \mathbb{T}^2 \to \mathbb{T}^2$ be a hyperbolic toral automorphism then there exists $\delta > 0$ such that any uniformly δ-close $S : \mathbb{T}^2 \to \mathbb{T}^2$ is semi-conjugate to T.*

PROOF. We use Theorem 5.4 to choose $\epsilon > 0$ sufficiently small that there exists $\delta > 0$ with every δ-pseudo-orbit being ϵ-close to the orbit of *exactly* one point in $\mathbb{T}^2$.

For any $x \in \mathbb{T}^2$ it follows from the definition of S being uniformly δ-close to T that the orbit $(x_n)_{n \in \mathbb{Z}} := (S^n x)_{n \in \mathbb{Z}}$ under S gives a δ-pseudo-orbit for T. By Theorem 5.4 there is a *unique* point $z \in \mathbb{T}^2$ whose T-orbit $(T^n(z))_{n \in \mathbb{Z}}$ is ϵ-close to $(x_n)_{n \in \mathbb{Z}}$ (i.e. $\sup_{n \in \mathbb{Z}} |T^n(z) - S^n(x)| \leq \epsilon$). We define $\pi : \mathbb{T}^2 \to \mathbb{T}^2$ by $\pi(x) := z$.

It is immediate that this is a semi-conjugacy, since from the definitions the orbit $(S^n(Sx))_{n \in \mathbb{Z}}$ of Sx (under S) is ϵ-close to the orbit of Tz (under T). By uniqueness, we see that $\pi(Sx) = Tz = T(\pi x)$.

We next show that $\pi : \mathbb{T}^2 \to \mathbb{T}^2$ is continuous. Let $\eta > 0$ be fixed and choose $N \geq 0$ such that $2(3\delta + 1)|\lambda_2|^N < \eta$. The continuity of $S : \mathbb{T}^2 \to \mathbb{T}^2$ shows that provided $x, y \in \mathbb{T}^2$ are sufficiently close then we have $|x_n - y_n| = |S^n(x) - S^n(y)| \leq \delta$ for $-N \leq n \leq N$ (where $x_n = S^n x$ and $y_n = S^n y$). In particular, $|T^n(\pi x) - T^n(\pi y)| \leq |T^n(\pi x) - x_n| + |x_n - y_n| + |y_n - T^n(\pi y)| \leq 3\delta$. However, if we write $\pi(x) = \pi(y) + \alpha v_1 + \beta v_2$ then $3\delta \geq |T^n(\pi x) - T^n(\pi y)| = |\alpha \lambda_1^n v_1 + \lambda_2^n \beta v_2|$. For $N \geq n \geq 1$ we see that $3\delta \geq |\lambda_1|^n \alpha - |\lambda_2|^n \beta$ and for $-N \leq n \leq -1$ we see that $3\delta \geq |\lambda_2|^n \beta - |\lambda_1|^n \alpha$. This allows us to see that $|\alpha|, |\beta| \leq (3\delta + 1)|\lambda_2|^N$ and so $|\pi(x) - \pi(y)| \leq |\alpha| + |\beta| \leq 2(3\delta + 1)|\lambda_2|^N < \eta$.

Finally, to see that $\pi : \mathbb{T}^2 \to \mathbb{T}^2$ is surjective we first observe from the proof of Theorem 5.4 that $S \to \pi_S := \pi$ is continuous and $\pi_T = I$ (where $I : \mathbb{T}^2 \to \mathbb{T}^2$ is the (surjective) identity map . Thus for S sufficiently close to T we have that π is surjective.

∎

COROLLARY 5.5.1. *Let $T : \mathbb{T}^2 \to \mathbb{T}^2$ be a hyperbolic toral automorphism; then there exists $\epsilon > 0$ such that for any uniformly ϵ-close homeomorphism $S : \mathbb{T}^2 \to \mathbb{T}^2$ we have $h(T) \leq h(S)$.*

PROOF. This follows by Theorem 5.4 and Proposition 3.5.

REMARK. If we make the strong assumption that $S : \mathbb{T}^2 \to \mathbb{T}^2$ is C^1 and S and T are uniformly close *and* their derivatives are close then T and S are conjugate. This property is called *structural stability* of $T : \mathbb{T}^2 \to \mathbb{T}^2$.

■

5.4 Comments and references

We have chosen to consider only hyperbolic toral automorphisms in two dimensions, rather than in arbitrary dimensions, to make the proofs more graphic. However, the results and proofs would be the same.

Good accounts of the general hyperbolic theory are in [3], [6], [5], etc. Other treatments of toral automorphisms are in [2] and [7]. For other important related topics see [2], [6], [7] (for structural stability and stable manifolds), [4], [1], [6, chapter 10], [3] (for Markov partitions)

References

1. R. Adler and B. Weiss, *Similarity of automorphisms of the torus*, Memoirs Amer. Math. Soc. **98** (1970).
2. V. Arnol'd, *Ordinary Differential Equations*, M.I.T. Press, Cambridge Mass., 1973.
3. R. Bowen, *Equilibrium States and the Ergodic Theory of Anosov Diffeomorphims*, Lecture Notes in Mathematics, 470, Springer, Berlin, 1975.
4. R. Bowen, *On Axiom A diffeomorphisms*, CMBS Reg. Conference series, 35, Amer. Math. Soc., Providence R.I., 1978.
5. R. Mañé, *Ergodic Theory and Differentiable Dynamics*, Springer, Berlin, 1987.
6. M. Shub, *Global Stability of Dynamical Systems*, Springer, Berlin, 1986.
7. W. Szlenk, *Introduction to the Theory of Smooth Dynamical Systems*, Wiley, New York, 1984.

CHAPTER 6

ROTATION NUMBERS

In this chapter we shall define the useful concept of the rotation number for orientation preserving homeomorphisms of the circle.

6.1 Homeomorphisms of the circle and rotation numbers

Let $T : \mathbb{R}/\mathbb{Z} \to \mathbb{R}/\mathbb{Z}$ be an orientation preserving homeomorphism of the circle to itself. There is a canonical projection $\pi : \mathbb{R} \to \mathbb{R}/\mathbb{Z}$ given by $\pi(x) = x \pmod 1$. We call a monotone map $\hat{T} : \mathbb{R} \to \mathbb{R}$ a *lift* of T if the canonical projection $\pi : \mathbb{R} \to \mathbb{R}/\mathbb{Z}$ is a semi-conjugacy (i.e. $\pi \circ \hat{T} = T \circ \pi$).

For a given map $T : \mathbb{R}/\mathbb{Z} \to \mathbb{R}/\mathbb{Z}$ a lift $\hat{T} : \mathbb{R} \to \mathbb{R}$ will not be unique.

EXAMPLE. If $T(x) = (x + \alpha) \pmod 1$ then for any $k \in \mathbb{Z}$ the map $\hat{T} : \mathbb{R} \to \mathbb{R}$ defined by $\hat{T}(x) = x + \alpha + k$ is a lift. To see this observe that $\pi(\hat{T}(x)) = \pi(x + \alpha + k) = x + \alpha \pmod 1$ and $T(\pi(x)) = \pi(x) + \alpha \pmod 1 = x + \alpha \pmod 1$.

The following lemma summarizes some simple properties of lifts.

LEMMA 6.1.

(i) *Let $T : \mathbb{R}/\mathbb{Z} \to \mathbb{R}/\mathbb{Z}$ be a homeomorphism of the circle; then if $\hat{T} : \mathbb{R} \to \mathbb{R}$ is a lift, then any other lift $\hat{T}' : \mathbb{R} \to \mathbb{R}$ must be of the form $\hat{T}'(x) = \hat{T}(x) + k$, for some $k \in \mathbb{Z}$.*

(ii) *For any $x, y \in \mathbb{R}$ with $|x - y| \leq k$ ($k \in \mathbb{Z}^+$) we have $|\hat{T}(x) - \hat{T}(y)| \leq k$. Iterating this gives that*

$$|\hat{T}^n(x) - \hat{T}^n(y)| \leq k, \quad \forall n \geq 0.$$

PROOF. These are easily seen from the continuity and the monotonicity of $\hat{T}$.

∎

DEFINITION. We define the *rotation number* $\rho(T)$ of the homeomorphism by

$$\rho(T) = \limsup_{n \to +\infty} \frac{\hat{T}^n(x)}{n} \pmod 1.$$

(The limsup is independent of the choice $x \in \mathbb{R}$ by Lemma 6.1. The choice of lift $\hat{T}$ can only alter the limsup by an integer, which has no bearing since we define $\rho(T)$ modulo one.)

EXAMPLE. Consider the standard rotation $R_\rho : \mathbb{R}/\mathbb{Z} \to \mathbb{R}/\mathbb{Z}$ defined by $R_\rho(x) = x + \rho \pmod 1$, where $\rho \in [0,1)$, say. Any lift $\hat{R}_\rho : \mathbb{R} \to \mathbb{R}$ will be of the form $\hat{R}_\rho(x) = x + \rho + k$, for some $k \in \mathbb{Z}$ i.e. translation on the real line by $\rho + k$. It is now immediate from the definition that the rotation number for R_ρ is merely $\rho \pmod 1$.

We can now show some interesting properties of $\rho(T)$.

PROPOSITION 6.2.

(i) *For $n \geq 1$ we have that $\rho(T^n) = n\rho(T) (\mathrm{mod}\ 1)$.*
(ii) *If T has a periodic point (i.e. $\exists n \geq 1, \exists x \in \mathbb{R}/\mathbb{Z}$ such that $T^n x = x$) then $\rho(T)$ is rational.*
(iii) *If $T : \mathbb{R}/\mathbb{Z} \to \mathbb{R}/\mathbb{Z}$ has no periodic points then $\rho(T)$ is irrational.*
(iv) *The limit actually exists and we can write*

$$\rho(T) = \limsup_{n \to +\infty} \frac{\hat{T}^n(x)}{n} (\mathrm{mod}\ 1).$$

PROOF. (i) Since $\hat{T}^n$ (the n th iterate of the lift $\hat{T}$ for T) is itself a lift for $T^n : \mathbb{R}/\mathbb{Z} \to \mathbb{R}/\mathbb{Z}$ this is immediate from the definitions.

(ii) Since $T^n(x + \mathbb{Z}) = x + \mathbb{Z}$ we have that $\hat{T}^n(x)$ and x differ by an integer (i.e. $\hat{T}^n(x) - x = k \in \mathbb{Z}$). Then for $pn + r$ with $0 \leq r \leq n - 1$ and $p \geq 0$ we have that $\hat{T}^{pn+r}(x) = \hat{T}^r(\hat{T}^{pn}x) = \hat{T}^r(x) + pk$ because of Lemma 6.1. Thus $\rho(T) = \limsup_{p \to +\infty} \frac{\hat{T}^{pn+r}(x)}{pn+r} = \frac{k}{n} \pmod 1$.

(iii) Assume for contradiction that $\rho(T) = \frac{p}{q}$ is rational. By part (i) we see that for $S := T^q$ we have $\rho(S) = 0$ and since T has no periodic points we conclude that $S : \mathbb{R}/\mathbb{Z} \to \mathbb{R}/\mathbb{Z}$ has no fixed points.

If $\hat{S} : \mathbb{R} \to \mathbb{R}$ is a lift for S then the absence of fixed points for S implies either $\hat{S}(x) > x, \forall x \in \mathbb{R}$, or $\hat{S}(x) < x, \forall x \in \mathbb{R}$. Assume $\hat{S}(x) > x, \forall x \in \mathbb{R}$ (the other case being similar), i.e. $\hat{S}$ is strictly increasing. If $\exists k > 0$ with $\hat{S}^k(0) > 1$ then we see that $\hat{S}^{mk}(0) > m$ and so

$$\rho(S) = \limsup_{n \to +\infty} \frac{\hat{S}^n(x)}{n} > \frac{1}{k},$$

contradicting that $\rho(S) = 0$. This leaves the possibility that $\hat{S}^k(0) < 1$ for all $k \geq 1$. Since $\hat{S}$ is strictly increasing the sequence $(\hat{S}^k(0))_{k=1}^{\infty}$ is monotone increasing and the supremum $z \in \mathbb{R}$ satisfies $\hat{S}(z) = z$. Thus S has a fixed point $S(z + \mathbb{Z}) = z + \mathbb{Z}$, giving a contradiction.

(iv) If T has a periodic point $T^n x = x$ then the argument in part (ii) actually shows that $\rho(T) = \lim_{N \to +\infty} \frac{T^N(x)}{N} = \frac{k}{n} \pmod 1$, in particular, showing that the limit exists.

Assume that $T : \mathbb{R}/\mathbb{Z} \to \mathbb{R}/\mathbb{Z}$ has no periodic points. Thus for all $n \geq 1$ there exists $k_n \in \mathbb{Z}$ such that $\hat{T}^n(x) - x \in [k_n, k_n + 1]$, $\forall x \in \mathbb{R}$, and, in particular, observe that $|\frac{\hat{T}^n(0)}{n} - \frac{k_n}{n}| < \frac{1}{n}$. Then, for any $m \geq 1$ we have that

$$
\begin{aligned}
\hat{T}^{nm}(0) = \hat{T}^n \left(\hat{T}^{n(m-1)}(0) \right) &- \left(\hat{T}^{n(m-1)}(0) \right) \\
+ \hat{T}^n \left(\hat{T}^{n(m-2)}(0) \right) &- \left(\hat{T}^{n(m-2)}(0) \right) + \ldots \\
\ldots + \hat{T}^n \left(T^n(0) \right) &- \left(\hat{T}^n(0) \right) + \hat{T}^n(0) \in [mk_n, m(k_n + 1)].
\end{aligned} \tag{6.1}
$$

In particular, we see from (6.1) that $|\frac{\hat{T}^{nm}(0)}{nm} - \frac{k_n}{n}| < \frac{1}{n}$. The triangle inequality gives that

$$
\begin{aligned}
|\frac{\hat{T}^m(0)}{m} - \frac{\hat{T}^n(0)}{n}| &\leq |\frac{\hat{T}^m(0)}{m} - \frac{k_m}{m}| + |\frac{k_m}{m} - \frac{\hat{T}^{mn}(0)}{mn}| \\
&+ |\frac{\hat{T}^{mn}(0)}{mn} - \frac{k_n}{n}| + |\frac{k_n}{n} - \frac{\hat{T}^n(0)}{n}| \\
&\leq \frac{2}{m} + \frac{2}{n}
\end{aligned}
$$

which shows that the sequence $\left(\frac{\hat{T}^n(0)}{n} \right)_{n=0}^{\infty}$ is Cauchy, and in particular the limit exists.   ∎

The next lemma and its corollary will be useful later.

LEMMA 6.3.

(i) *Let $n_1, n_2, m_1, m_2 \in \mathbb{Z}$ and $x, y \in \mathbb{R}$. If $\hat{T}^{n_1}(x) + m_1 < \hat{T}^{n_2}(x) + m_2$ then $\hat{T}^{n_1}(y) + m_1 < \hat{T}^{n_2}(y) + m_2$;*
(ii) *The bijection $n\rho(T) + m \to \hat{T}^n(0) + m$ between the sets*

$$\Omega = \{n\rho(T) + m : n, m \in \mathbb{Z}\} \text{ and } \Lambda = \{\hat{T}^n(0) + m : n, m \in \mathbb{Z}\}$$

preserves the natural ordering on $\mathbb{R}$.

PROOF. (i) If $\exists x, y \in \mathbb{R}$ for which the ordering is reversed, then by continuity (and the intermediate value theorem) there exists $z \in \mathbb{R}$ with $\hat{T}^{n_1}(z) + m_1 = \hat{T}^{n_2}(z) + m_2$, i.e. $\hat{T}^{n_1}(z) - \hat{T}^{n_2}(z) \in \mathbb{Z}$. But then $T^{n_1 - n_2}(z + \mathbb{Z}) = z + \mathbb{Z}$ is a periodic point. This contradicts the assumption that T has irrational rotation number (and so no periodic points by Proposition 6.2 (iii)).

(ii) Assume that $\hat{T}^{n_1}(0) + m_1 < \hat{T}^{n_2}(0) + m_2$; then we wish to show that $n_1\rho + m_1 < n_2\rho + m_2$. We can rewrite the first inequality as $\hat{T}^{n_1-n_2}(T^{n_2}0) - T^{n_2}0 < (m_2 - m_1)$ and we can apply part (i) with $x = T^{n_2}0$ and $y = 0$ to deduce that

$$\hat{T}^{n_1-n_2}(0) < (m_2 - m_1). \tag{6.2}$$

Next, we can apply part (i) to (6.2) with the choices $x = \hat{T}^{n_1-n_2}(0)$ and $y = 0$ to deduce

$$\hat{T}^{2(n_1-n_2)}(0) - \hat{T}^{n_1-n_2}(0) = \hat{T}^{n_1-n_2}(\hat{T}^{n_1-n_2}0) - \hat{T}^{n_1-n_2}(0) \\ < (m_2 - m_1). \tag{6.3}$$

Comparing (6.2) and (6.3) gives that $\hat{T}^{2(n_1-n_2)}(0) < 2(m_2 - m_1)$. Proceeding inductively shows that for any $N \geq 1$ we have $\hat{T}^{N(n_1-n_2)}(0) < N(m_2 - m_1)$. Finally, we see that

$$\rho(T) = \lim_{n \to +\infty} \frac{\hat{T}^n(0)}{n} = \lim_{N \to +\infty} \frac{\hat{T}^{N(n_1-n_2)}(0)}{N(n_1 - n_2)} \leq \frac{(m_2 - m_1)}{(n_1 - n_2)}$$

and in fact $\rho < \frac{(m_2-m_1)}{(n_1-n_2)}$ since ρ is assumed irrational. This is the required inequality, and so this completes the proof. ∎

COROLLARY 6.3.1. *Let $T : \mathbb{R}/\mathbb{Z} \to \mathbb{R}/\mathbb{Z}$ have irrational rotation number ρ. For any $x \in \mathbb{R}/\mathbb{Z}$ the orbits of x under T and the rotation $R_\rho : \mathbb{R}/\mathbb{Z} \to \mathbb{R}/\mathbb{Z}$ have the same ordering.*

PROOF. This follows immediately from part (ii) of Lemma 6.3, since a difference in the ordering of $T^n(x)$ and $R_\rho^n(x)$ would contradict the conclusion of the lemma.

∎

6.2 Denjoy's theorem

The following result gives a sufficient condition for a homeomorphism to be conjugate to a rotation.

PROPOSITION 6.4. *If $T : \mathbb{R}/\mathbb{Z} \to \mathbb{R}/\mathbb{Z}$ is a minimal orientation preserving homeomorphism with irrational rotation number ρ then T is topologically conjugate to the standard rotation $R_\rho : \mathbb{R}/\mathbb{Z} \to \mathbb{R}/\mathbb{Z}$.*

PROOF. Let $\hat{T} : \mathbb{R} \to \mathbb{R}$ be a lift of $T : \mathbb{R}/\mathbb{Z} \to \mathbb{R}/\mathbb{Z}$. Observe that since ρ is irrational we have that $\Omega \subset \mathbb{R}$ (as defined in Lemma 6.3 (ii)) is dense. Moreover, since T is minimal we know that $\{T^n 0\}$ is dense in $\mathbb{R}/\mathbb{Z}$, and so we also have that $\Lambda \subset \mathbb{R}$ is dense.

The map $\phi : \Lambda \to \Omega$ given by $\phi(\hat{T}^n(0) + m) = n\rho + m$ is order preserving by Lemma 6.3. Thus it extends to a homeomorphism $\phi : \mathbb{R} \to \mathbb{R}$.

Observe that

$$\phi\left(\hat{T}(\hat{T}^n(0) + m)\right) = \phi\left(\hat{T}^{n+1}(0) + m\right) = (n+1)\rho + m$$

and

$$\hat{R}_\rho \phi(\hat{T}^n(0) + m) = \hat{R}_\rho(n\rho + m) = (n+1)\rho + m$$

where $\hat{R}_\rho : \mathbb{R} \to \mathbb{R}$ is a lift for R_ρ. Thus $\phi \circ \hat{T} = \hat{R}_\rho \circ \phi$.

Finally, we observe that by construction $\phi(x+1) = \phi(x) + 1$. Thus the homeomorphism $\phi : \mathbb{R}/\mathbb{Z} \to \mathbb{R}/\mathbb{Z}$ defined by $\phi(x + \mathbb{Z}) = \phi(x) + \mathbb{Z}$ is well-defined. Moreover, the identity $\phi \circ \hat{T} = \hat{R}_\rho \circ \phi$ implies the conjugacy relation $\phi \circ T = R_\rho \circ \phi$. This completes the proof of the proposition.

∎

Given a C^1 map $T : \mathbb{R}/\mathbb{Z} \to \mathbb{R}/\mathbb{Z}$ we consider its derivative $T' : \mathbb{R}/\mathbb{Z} \to \mathbb{R}$.

DEFINITION. We define the *variation* of $\log|T'| : \mathbb{R}/\mathbb{Z} \to \mathbb{R}$ by

$$\mathrm{Var}(\log|T'|)$$
$$= \sup\left\{ \sum_{i=0}^{n-1} |\log|T'|(x_{i+1}) - \log|T'|(x_i)| : 0 = x_0 < x_1 < \ldots < x_n = 1 \right\}.$$

We say that the logarithm of $|T'|$ has *bounded variation* if this value $\mathrm{Var}(\log|T'|)$ is finite.

It is easy to see that if $T : \mathbb{R}/\mathbb{Z} \to \mathbb{R}/\mathbb{Z}$ is C^2 then $\mathrm{Var}(\log|T'|)$ is finite.

We now come to the main result of this section which gives sufficient conditions for a homeomorphism to be conjugate to a rotation (and are more readily checked than those of Proposition 6.4).

THEOREM 6.5 (DENJOY'S THEOREM). *If* $T : \mathbb{R}/\mathbb{Z} \to \mathbb{R}/\mathbb{Z}$ *is a* C^1 *orientation preserving homeomorphim of the circle with derivative of bounded variation and irrational rotation number* $\rho = \rho(T)$ *then* $T : \mathbb{R}/\mathbb{Z} \to \mathbb{R}/\mathbb{Z}$ *is topologically conjugate to the standard rotation* $R_\rho : \mathbb{R}/\mathbb{Z} \to \mathbb{R}/\mathbb{Z}$.

PROOF. It suffices to show that T is minimal, then the result follows by applying Proposition 6.4. The proof of minimality will come via two sublemmas.

SUBLEMMA 6.5.1. *If T has irrational rotation number and there are a constant $C > 0$ and a sequence of integers $q_n \to +\infty$ such that the maps $T^{q_n} : \mathbb{R}/\mathbb{Z} \to \mathbb{R}/\mathbb{Z}$ satisfy*

$$|(T^{q_n})'(x)| \cdot |(T^{-q_n})'(x)| \leq C$$

then $T : \mathbb{R}/\mathbb{Z} \to \mathbb{R}/\mathbb{Z}$ is minimal

PROOF. If T is not minimal we may choose $x \in \mathbb{R}/\mathbb{Z}$ such that $Y = \mathrm{cl}(\cup_{n \in \mathbb{Z}} T^n x) \neq X$. We can choose a (maximal) interval $I_0 \subset X - Y$; then we claim that $I_n := T^{-n} I_0 \subset X - Y$ are distinct (maximal) intervals. To see this we observe that by maximality I_0 must be of the form $I_0 = (a, b)$ (with $a, b \in Y$). Thus $I_n = (T^{-n} a, T^{-n} b)$ and if $I_n \cap I_m \neq \emptyset$ then again by maximality $I_n = I_m$ and, in particular, $T^{-n} a = T^{-m} a$. But if $n \neq m$ then this means a is a periodic point, which contradicts T having an irrational rotation number (by Lemma 6.1 (ii)).

If $|I_n|$ denotes the length of the interval I_n, $n \in \mathbb{Z}$, then by the disjointness we see that $\sum_{n \in \mathbb{Z}} |I_n| \leq 1$. In particular, $|I_n| \to 0$ as $n \to +\infty$.

However, we see that for all $n \geq 1$,

$$|I_{q_n}| + |I_{q_{-n}}| = \int_{I_0} \left(|(T^{q_n})'(x)| + |(T^{-q_n})'(x)| \right) dx$$

$$\geq 2 \int_{I_0} \left(|(T^{q_n})'(x)||(T^{-q_n})'(x)| \right)^{\frac{1}{2}} dx$$

$$\geq 2C^{\frac{1}{2}} |I_0|$$

(since arithmetic averages are larger than geometric averages). This contradicts $|I_n| \to 0$, and so completes the proof of the sublemma. ∎

To make use of the assumption of the bounded variation of $\log |T'|$ we need a second sublemma.

SUBLEMMA 6.5.2. *Fix $x \in \mathbb{R}/\mathbb{Z}$ and wrote $x_n = T^n(x)$, for $n \in \mathbb{Z}$. There exists an increasing sequence $q_n \to +\infty$ of natural numbers such that the intervals*

$$(x_0, x_{q_n}), (x_1, x_{q_n+1}), (x_2, x_{q_n+2}), \ldots, (x_i, x_{q_n+i}), \ldots, (x_{q_n}, x_{2q_n})$$

are all disjoint.

PROOF.. By Corollary 6.3.1 we see that the order on $\mathbb{R}/\mathbb{Z}$ of points in the orbit of $T : \mathbb{R}/\mathbb{Z} \to \mathbb{R}/\mathbb{Z}$ is the same as that of the rotation $R_\rho : \mathbb{R}/\mathbb{Z} \to \mathbb{R}/\mathbb{Z}$. Thus it suffices to prove this sublemma with R_ρ rather than T.

For each $n \geq 1$ we want to choose the sequence q_n to correspond to the successive nearest approaches of $\{T^m x\}$ to x (i.e. $|T^{q_n} x - x| < \delta_n =$

$\inf\{|T^j x - x| : 1 \le j \le q_n - 1\}$). Consider a typical interval (x_i, x_{q_n+i}) $(0 \le i \le q_n)$ and assume for a contradiction that there exists $x_r \in (x_i, x_{q_n+i})$ $(0 \le r \le 2q_n)$. There are two cases.

(a) Firstly, assume that $r < i$. We then know that

$$x_0 = R_\rho^{-r}(x_r) \in R_\rho^{-r}(x_i, x_{q_n+i}) = (x_{(i-r)}, x_{q_n+(i-r)}).$$

In particular, $|x_{i-r} - x_0| < |x_{q_n+(i-r)} - x_{(i-r)}| = |x_{q_n} - x_0| = \delta_n$. But since $q_n > (i - r) > 0$ this contradicts the definition of q_n.

(b) Secondly, assume that $r > i$. We then know that

$$x_{r-i} = R_\rho^{-i}(x_r) \in R_\rho^{-i}(x_i, x_{q_n+i}) = (x_0, x_{q_n})$$

and so we see that $r - i > q_n$ (from the definition of q_n). But then $x_{(r-i)-q_n} = R_\rho^{-q_n}(x_{r-i}) \in R_\rho^{-q_n}(x_0, x_{q_n}) = (x_{-q_n}, x_0)$ and, in particular, $|x_{(r-i)-q_n} - x_0| < |x_{-q_n} - x_0| = |R_\rho^{q_n}(x_{-q_n}) - R_\rho^{q_n}(x_0)| = |x_0 - x_{q_n}| = \delta_n$. Since $0 < (r - i) - q_n < q_n$ this contradicts the definition of q_n.

This completes the proof of the sublemma. ∎

Since the intervals in Sublemma 6.5.2 are disjoint we have for any $n \ge 1$

$$\begin{aligned}
\mathrm{Var}(\log|T'|) &\ge \sum_{i=0}^{q_n} |\log|T'(x_i)| - \log|T'(x_{q_n+i})|| \\
&= \left| \sum_{i=0}^{q_n} \log|T'(x_i)| - \sum_{i=0}^{q_n} \log|T'(x_{q_n+i})| \right| \\
&= \left| \log\left(\frac{\prod_{i=0}^{q_n-1} |T'(T^i x_0)|}{\prod_{i=0}^{q_n-1} |T'(T^i x_{q_n})|} \right) \right| \\
&= \left| \log\left(\frac{|(T^{q_n})'(x_0)|}{|(T^{q_n})'(x_{q_n})|} \right) \right| \\
&= \left| \log|(T^{q_n})'(x_{q_n})(T^{-q_n})'(x_{q_n})| \right|
\end{aligned} \tag{6.4}$$

(where by the chain rule $(T^{-q_n})'(x_{q_n})(T^{q_n})'(x_0) = 1$). Since this holds for arbitrary x the point x_{q_n} can be replaced by an arbitrary point on the circle. If we take the exponential in identity (6.4) then Theorem 6.5 now follows from Sublemma 6.5.1 and 6.4. ∎

REMARK. We should remark that the assumption that $\log T'$ has bounded variation is necessary. If we relax this assumption then we have that T may be merely semi-conjugate to the rotation R_ρ.

6.3 Comments and references

Some basic results about rotation numbers can be found in [2, pp. 102-108], [5, chapter 12] and [1].

The question of when the conjugating map is differentiable is subtler (cf.[4], [3], [9] and [6], [7], [8] for variants).

References

1. V. Arnol'd, *Ordinary Differential Equations*, M.I.T. Press, Cambridge Mass., 1973.
2. R. Devaney, *An Introduction to Chaotic Dynamical Systems*, Addison-Wesley, New York, 1989.
3. P. Deligne, *Les difféomorphismes du cércle (d'après M. R. Herman)*, Exposé 477, Semin. Bourbaki 1975/76, Lect. Notes Math. 567, Springer, Berlin, 1977.
4. M. Herman, *Sur la conjugason différentiable des diffeomporhismes du cercle à des rotations*, Publ. Math. (IHES) **49** (1979), 5-234.
5. A. Katok and B. Hasselblatt, *An Introduction to the Modern Theory of Dynamical Systems*, C.U.P., Cambridge, 1995.
6. Y. Katznelson and D. Ornstein, *The differentiability of the conjugation of certain diffeomorphisms of the circle*, Ergod. Th. and Dynam. Sys. **9** (1989), 643-680.
7. K. Khanin and Y. Sinai, *A new proof of M. Herman's theorem*, Commun. Math. Phys. **112** (1987), 89-101.
8. D. Rand, *Universality and renormalisation in dynamical systems*, New Directions in Dynamical Systems (T. Bedford and J. Swift, ed.), C.U.P., Cambridge, 1988, pp. 1-56.
9. J.-C. Yoccoz, *Conjugaison différentable des difféomorphismes du cercle dont le nombre de rotation vérifie une condition Diophantienne*, Ann. Sci. Ec. Norm. Sup. **17** (1984), 333-361.

CHAPTER 7

INVARIANT MEASURES

In this chapter we shall introduce some basic definitions in ergodic theory.

7.1 Definitions and characterization of invariant measures

Let $(X, \mathcal{B}, \mu)$ be a measure space. Assume that μ is a probability measure, i.e. $\mu(X) = 1$.

DEFINITION. A measurable map $T : X \to X$ (i.e. $T^{-1}\mathcal{B} \subset \mathcal{B}$) is said to preserve the measure μ if for any $B \in \mathcal{B}$ we have $\mu(B) = \mu(T^{-1}B)$. Alternatively we say that μ is T-invariant.

The next lemma gives a characterization in terms of integrable functions.

LEMMA 7.1. *T preserves μ iff $\int f d\mu = \int f \circ T d\mu$ for all $f \in L^1(X, \mathcal{B}, \mu)$.*

PROOF. This is easy, using indicator functions.

∎

7.2 Borel sigma-algebras for compact metric spaces

We now specialize to homeomorphisms $T : X \to X$ of a compact metric space X. In this case a natural sigma-algebra is the Borel sigma-algebra $\mathcal{C}$, i.e. the smallest sigma-algebra containing all of the open sets in X.

LEMMA 7.2. *T preserves μ iff $\int f d\mu = \int f \circ T d\mu$ for all $f \in C^0(X)$.*

PROOF. One way follows from Lemma 7.1. The other follows from the Hahn-Banach and Riesz representation theorems.

∎

The next proposition shows that there always exists an invariant measure.

PROPOSITION 7.3 (EXISTENCE OF INVARIANT MEASURES). *Let X be a compact metric space and $\mathcal{B}$ be the Borel sigma-algebra. Given any homeomorphism $T : X \to X$ (or more generally, a continuous map) there exists at least one probability measure μ preserving T.*

PROOF. Choose

(1) a countable dense subset $\{ f_k \in C^0(X, \mathbb{R}) : k \geq 0 \}$, in the uniform topology
(2) a point $x \in X$.

We can consider the averages $\frac{1}{N}\sum_{n=0}^{N-1} f_k(T^n x) \in \mathbb{R}$, for $k \geq 0, N \geq 1$; then clearly $-||f_k||_\infty \leq \frac{1}{N}\sum_{n=0}^{N-1} f_k(T^n x) \leq ||f_k||_\infty$. By compactness of the interval $[-||f_0||_\infty, ||f_0||_\infty]$ we can choose a sub-sequence $\mathcal{N}^{(0)} = \{N_r^{(0)}\}_{r=0}^\infty \subset \mathbb{Z}^+$ such that

$$\frac{1}{N_r^{(0)}} \sum_{n=0}^{N_r^{(0)}-1} f_0(T^n x) \to C(f_0) \text{ as } r \to +\infty.$$

We repeat the argument with f_1 replacing f_0. By compactness of the interval $[-||f_1||_\infty, ||f_1||_\infty]$ we can choose a subsequence $\mathcal{N}^{(1)} = \{N_r^{(1)}\}_{r=0}^\infty \subset \mathcal{N}^{(0)}$ such that $\frac{1}{N_r^{(1)}} \sum_{n=0}^{N_r^{(1)}-1} f_1(T^n x) \to C(f_1)$ as $r \to +\infty$.

Repeating the argument inductively we can arrange subsequences $\mathcal{N}^{(m)} = \{N_r^{(m)}\}_{r=0}^\infty$ with $\mathcal{N}^{(m)} \subset \mathcal{N}^{(m-1)} \subset \ldots \subset \mathcal{N}^{(1)} \subset \mathcal{N}^{(0)}$ for $m \geq 0$ such that

$$\frac{1}{N_r^{(m)}} \sum_{n=0}^{N_r^{(m)}-1} f_m(T^n x) \to C(f_m)$$

as $r \to +\infty$.

Next we choose a diagonal sub-sequence $N_n = N_n^{(n)}$ and we conclude that for *any* $k \geq 0$ we have that

$$\frac{1}{N_n^{(n)}} \sum_{n=0}^{N_n^{(n)}-1} f_k(T^n x) \to C(f_k)$$

as $n \to +\infty$.

We claim that for any $f \in C^0(X)$ we have that $\frac{1}{N_n^{(n)}}\sum_{n=0}^{N_n^{(n)}-1} f(T^n x)$, $n \geq 1$, converges (to a limit we shall call $C(f)$) as $n \to +\infty$.

For any $\epsilon > 0$ we choose $||f - f_k||_\infty < \epsilon$ and thus

$$\limsup_{n\to+\infty} \left| \frac{1}{N_n^{(n)}} \sum_{n=0}^{N_n^{(n)}-1} f(T^n) - C(f_k) \right| \leq \epsilon.$$

Since $\epsilon > 0$ can be chosen arbitrarily small this is enough to show that the sequence $\frac{1}{N_n^{(n)}}\sum_{n=0}^{N_n^{(n)}-1} f(T^n x)$ converges.

We can summarize the properties of the limit $C(f)$ as follows:

(1) $\forall f_1, f_2 \in C^0(X)$ and $\lambda_1, \lambda_2 \in \mathbb{R}$ we have that $C(\lambda_1 f_1 + \lambda_2 f_2) = \lambda_1 C(f_1) + \lambda_2 C(f_2)$ and $|C(f)| \leq ||f||_\infty$,

(2) if $f(x) \geq 0$ $\forall x \in X$ we have $C(f) \geq 0$,

(3) $C(1) = 1$, where $1(x) = 1$, $\forall x \in X$,

(4) $C(f \circ T) = C(f)$ (since it is true for each $f = f_k$ by

$$C(f_k \circ T) = \lim_{n \to +\infty} \frac{1}{N_n^{(n)} - 1} \sum_{n=0}^{N_n^{(n)}} f_k(T^{n+1}x)$$

$$= \lim_{n \to +\infty} \frac{1}{N_n^{(n)}} \left(f_k(x) + \sum_{n=0}^{N_n^{(n)}-1} f_k(T^n x) - f_k(T^{N_n^{(n)}} x) \right) = C(f_k)$$

and continuity).

Using the Riesz representation theorem, we see that the linear functional $C : C^0(X) \to \mathbb{R}$ corresponds to a probability measure μ. The last property implies that $\int fT d\mu = \int f d\mu$ for all $f \in C^0(X)$, which is equivalent to T-invariance.   ∎

7.3 Examples of invariant measures

It is illuminating to look at some simple examples.

EXAMPLE 1 (ROTATIONS ON TORI). Let $X = \mathbb{T}^n = \mathbb{R}^n / \mathbb{Z}^n$ be a torus. Define a generalized rotation $T : X \to X$ (or *Kronecker system*) by fixing a vector $(a_1, \dots, a_n)$ and defining

$$T\left((x_1, \dots, x_n) + \mathbb{Z}^n\right) = (x_1 + a_1, \dots, x_n + a_n) + \mathbb{Z}^n.$$

We let $\mathcal{B}$ be the Borel sigma-algebra. We let μ be the Lebesgue-Haar measure and then naturally μ is T-invariant. When $n = 1$ this is simply rotation on a circle.

One way to see this is the following. For any $f(x_1, \dots, x_n) \in C^0(X)$ we write it as a Fourier series:

$$f(x_1, \dots, x_n) = \sum_{k_1, \dots, k_n = -\infty}^{\infty} a_{k_1, \dots, k_n} e^{2\pi i (k_1 x_1 + \dots + k_n x_n)}.$$

Moreover, we see that $a_0 = \int f dx_1 \dots dx_n$ (just by integrating both sides). The action of the rotation can be written as

$$(fT)(x_1, \dots, x_n) = \sum_{k_1, \dots, k_n = -\infty}^{\infty} a_{k_1, \dots, k_n} e^{2\pi i (k_1(x_1 + a_1) + \dots + k_n(x_n + a_n))}$$

$$= \sum_{k_1, \dots, k_n = -\infty}^{\infty} b_{k_1, \dots, k_n} e^{2\pi i (k_1 x_1 + \dots + k_n x_n)}$$

where $b_0 = \int fT dx_1 \ldots dx_n$, as before. But by comparing terms we see that $b_0 = a_0$.

REMARK. More generally this is a property of Haar measure: Assume that X is a (locally) compact second countable metric space which is also a group with identity element e, where multiplication between $x, y \in X$ is denoted by xy and the inverse to x is denoted by x^{-1}. Assume that the map $X \times X \to X$ given by $(x, y) \to xy$ is continuous. By a general theorem (Haar measure theorem) there exists a (sigma) finite measure μ on the Borel sigma-algebra for X such that

(1) $\mu(U) > 0$ for every open set $U \subset X$,
(2) for any Borel set A and $g \in X$ we have $\mu(gA) = \mu(A)$.

More generally, we can take X to be a compact topological group and for any fixed element $g \in X$ we define a transformation $T : X \to X$ by $Tx = gx$. The (normalized) Haar measure is then an invariant probability measure.

EXAMPLE 2 (ADDING MACHINE). A slightly more exotic example is the adding machine transformation defined on $X = \prod_{n=0}^{\infty} \{0, 1\}$. This is a topological group under "addition with infinite carry". In particular, let $g = (1, 0, 0, 0, \ldots)$ and define $T : X \to X$ as follows: If $x = (0, 0, \ldots, 0, 1, a_k, a_{k+1}, a_{k+2}, \ldots)$ then $Tx = x + g = (1, 0, \ldots, 0, 1, a_k, a_{k+1}, a_{k+2}, \ldots)$, and if $x = (1, 1, \ldots, 1, 0, a_k, a_{k+1}, a_{k+2}, \ldots)$ then $Tx = x + g = (0, 0, \ldots, 0, 1, a_k, a_{k+1}, a_{k+2}, \ldots)$. The invariant measure here is given by the measure on cylinders,

$$[z_0, \ldots, z_n] = \{x \in X_A : x_i = z_i \text{ for } 0 \leq i \leq n\}$$

by $\mu([z_0, \ldots, z_n]) = \left(\frac{1}{2}\right)^n$.

EXAMPLE 3 (DOUBLING MAP). Let $T : \mathbb{R}/\mathbb{Z} \to \mathbb{R}/\mathbb{Z}$ be defined by $T(x) = 2x \pmod{1}$. Let $\mathcal{B}$ be the usual Borel sigma-algebra. Let μ be the usual Lebesgue-Haar measure. To show that μ is invariant, it suffices to show that for each interval $I = [a, b]$ we have that $\mu(T^{-1}I) = \mu(I)$. Writing

$$T^{-1}I = \left[\frac{a}{2}, \frac{b}{2}\right] \cup \left[\frac{a}{2} + \frac{1}{2}, \frac{b}{2} + \frac{1}{2}\right],$$

this becomes apparent.

TECHNICAL REMARK. The reason that it suffices to prove this only on intervals (or more generally on a basis for the topology) and not necessarily all sets is by virtue of the Kolmogorov extension theorem.

EXAMPLE 4 (MARKOV MEASURES). Let A be a $k \times k$ matrix with entries either 0 or 1, and let $\sigma : X_A \to X_A$ be the associated subshift of finite type. Let $P = \{P_{ij}\}$ be any $k \times k$ stochastic matrix, i.e.

(1) $P_{ij} \geq 0$,
(2) The row sums are unity (i.e. $\sum_{j=1}^{k} P_{ij} = 1$).

In addition we ask that $P_{ij} = 0$ whenever $A(i,j) = 0$. The second condition means that $(1, \ldots, 1)$ is a right eigenvector. We let $p = (p_1, \ldots, p_k)$ be the left eigenvector with eigenvalue 1. We can define a measure μ on X_A on the cylinders by

$$\mu([z_{-m}, \ldots, z_n]) = p_{z_{-m}} P_{z_{-m}, z_{-m+1}} P_{z_{-m+1}, z_{-m+2}} \cdots P_{z_{n-1}, z_n}.$$

This is enough to define the measure on the entire sigma-algebra (i.e. Kolmogorov extension theorem). Clearly the measure is shift-invariant. The use of p in the definition is to ensure that it is consistent, i.e.

$$\mu([z_{-m+1}, \ldots, z_n])$$
$$= p_{z_{-m+1}} P_{z_{-m+1}, z_{-m+2}} P_{z_{-m+1}, z_{-m+2}} \cdots P_{z_{n-1}, z_n}$$
$$= \left(\sum_{\substack{1 \le z_{-m} \le k \\ A(z_{-m}, z_{m-1})=1}} \frac{p_{z_{-m}}}{p_{z_{-m+1}}} P_{z_{-m}, z_{-m+1}} \right) p_{z_{-m+1}} P_{z_{-m+1}, z_{-m+2}} \cdots P_{z_{n-1}, z_n}$$
$$= \sum_{\substack{1 \le z_{-m} \le k \\ A(z_{-m}, z_{m-1})=1}} \mu([z_{-m}, \ldots, z_n])$$

This is called a *Markov measure*.

SPECIAL CASE: FULL SHIFT. In the special case that *all* of the entries in the matrix are 1s the shift $X_A = \prod_{n \in \mathbb{Z}} \{1, \ldots, k\}$ is called a full shift on k symbols. In this case let $p = (p_1, \ldots, p_k)$ be any probability vector (i.e. $p_1, \ldots, p_k \ge 0$ and $p_1 + \ldots + p_k = 1$). We then define $\mu([z_{-m}, \ldots, z_n]) = p_{z_{-m}} \cdots p_{z_n}$. This is called a Bernoulli measure.

ANOTHER EXAMPLE. In section 12.2 one can find another example involving expanding maps of the interval.

7.4 Invariant measures for other actions

In this section we shall consider the case where the discrete transformation T is replaced by either a flow (an $\mathbb{R}$-action) or a $\mathbb{Z}^n$-action. Although this will not play an important role in the sequel, we shall include a few comments here for completeness.

DEFINITION (INVARIANT MEASURES FOR FLOWS). A flow $T_t : X \to X$ is a one-parameter family of measurable invertible maps (i.e. $\forall t \in \mathbb{R}$ we have that for any $B \in \mathcal{B}$ we know $T_t^{-1} B \in \mathcal{B}$) which satisfy

(1) $T_t T_s = T_{s+t}$ for all $s, t \in \mathbb{R}$,
(2) T_0 is the identity map.

DEFINITION. A measurable flow $T_t : X \to X$ is said to preserve the measure μ if for any $B \in \mathcal{B}$ we have $\mu(B) = \mu(T_t^{-1}B)$ for all $t \in \mathbb{R}$.

Alternatively we say that μ is T-invariant.

EXAMPLE 5 (SUSPENDED FLOWS). Consider a measure space $(X, \mathcal{B}, \mu)$ and a transformation $T : X \to X$ which preserves the measure μ. Given any function $r : X \to \mathbb{R}^+$ with $\int r d\mu < +\infty$ we define a new space by

$$X^r = \{(x,t) \in X \times \mathbb{R} : 0 \leq t \leq r(x)\}$$

where we identify the points $(x, r(x)) = (Tx, 0)$. We associate to this the product sigma algebra (from the sigma-algebra on X and the sigma-algebra on the real line). We define a (probability) measure μ^r on X^r by $d\mu^r = \frac{1}{\int r d\mu} d\mu \times dt$. (Here the factor $\frac{1}{\int r d\mu}$ appears to normalize this to be a probability measure.)

We define a flow $T_t : X^r \to X^r$ by

$$T_t(x, u) = (T^{n-1}x, u + t - \sum_{i=0}^{n-1} r(T^i x))$$

whenever $t \geq 0$ and $0 \leq u + t - \sum_{i=0}^{n-1} r(T^i x) \leq r(T^n x)$, and

$$T_t(x, u) = (T^{-n}x, u + t + \sum_{i=-n}^{-1} r(T^i x))$$

whenever $t \leq 0$ and $0 \leq u + t + \sum_{i=-n}^{-1} r(T^{-i}x) \leq r(T^{-n+1}x)$. It is easy to see that this is a measurable flow, and that the measure μ^r is invariant.

EXAMPLE 6 (GEODESIC AND HOROCYCLE FLOWS). Consider the group

$$G = \{ \begin{pmatrix} a & b \\ c & d \end{pmatrix} : a, b, c, d \in \mathbb{R}, \qquad ad - bc = 1 \}$$

This is a *locally* compact topological group. There is an associated Haar measure defined by $d\mu(g) = \frac{1}{|d|}(da)(db)(dc)$ on G (although the total measure of the space G is infinite!).

Assume that $\Gamma \subset G$ is a discrete subgroup such that the measure of the quotient space G/Γ is finite (i.e. we can find $B \subset G$ such that $B/\Gamma = G/\Gamma$ and $\mu(B) < +\infty$).

We can define two flows on G/Γ as follows:

(1) the geodesic flow $\phi_t : g\Gamma \to g_t g\Gamma$ defined by left multiplication by the matrices $g_t = \begin{pmatrix} e^t & 0 \\ 0 & e^{-t} \end{pmatrix}$;

(2) the horocycle flow $\psi_t : g\Gamma \to h_t g\Gamma$ defined by left multiplication by the matrices $h_t = \begin{pmatrix} 1 & t \\ 0 & 1 \end{pmatrix}$.

Each of these flows preserves the measure on G/Γ corresponding to μ.

REMARK. Geometrically these correspond to flows on the unit tangent bundle of compact surfaces of constant negative curvature. The invariant measure then corresponds to the Liouville measure.

Let $T_1, \ldots, T_k : X \to X$ be a family of commuting measurable (invertible) transformations on a measurable space $(X, \mathcal{B})$. We can define a $\mathbb{Z}^k$-*action* $A : \mathbb{Z}^k \times X \to X$ by $A(n_1, \ldots, n_k; x) = T^{n_1} \circ \ldots \circ T^{n_k} x$.

DEFINITION. We say that a measure μ is invariant under the action A if for every $B \in \mathcal{B}$ we have that $\mu(B) = \mu(T_1^{-n_1} \ldots T_k^{-n_k} B)$ for all $(n_1, \ldots, n_k) \in \mathbb{Z}^n$ (or equivalently, $\mu(B) = \mu(T_i^{-1} B)$ for all $i = 1, \ldots, k$).

7.5 Comments and references

More examples of invariant measures for different transformations can be found in [2] and [4], and also particularly in [3].

Details on the theory of suspended flows can be found in [1, pp. 292-295].

References

1. I. Cornfold, S. Fomin and Y. Sinai, *Ergodic Theory*, Springer, Berlin, 1982.
2. A. Katok and B. Hasselblatt, *An Introduction to the Modern Theory of Dynamical Systems*, C.U.P., Cambridge, 1995.
3. F. Schweiger, *The Ergodic Theory of Fibred Systems and Metric Number Theory*, O.U.P., Oxford, 1994.
4. P. Walters, *An Introduction to Ergodic Theory*, Springer, Berlin, 1982.

CHAPTER 8

MEASURE THEORETIC ENTROPY

In this chapter we shall show how to associate to a measure preserving transformation an important quantity called the measure theoretic entropy. This gives important information on the dynamics of the map (cf. Chapter 12) and is useful in classifying measure preserving transformations.

The essential results are contained in sections 8.1-8.3. If one accepts Sinai's result on strong generators (Lemma 8.8) without proof then sections 8.4-8.8 will only be required again in chapter 12.

8.1 Partitions and conditional expectations

Let $\alpha = \{A_i\}_{i \in I}$ be a countable measurable partition of the probability space $(X, \mathcal{B}, \mu)$, i.e.

(i) $X = \cup_i A_i$ (up to a set of zero μ-measure), and
(ii) $A_i \cap A_j = \emptyset$ for $i \neq j$ (up to a set of zero μ-measure).

DEFINITION. We define the information function $I(\alpha) : X \to \mathbb{R}$ by

$$I(\alpha)(x) = -\sum_i \log \mu(A_i) \chi_{A_i}(x),$$

i.e. $I(\alpha)(x) = -\log \mu(A_i)$ if $x \in A_i$.

Consider a *sub*-sigma-algebra $\mathcal{A} \subset \mathcal{B}$; then we can define a measure space $(X, \mathcal{A}, \mu)$ with respect to the smaller sigma-algebra. For any $f \in L^1(X, \mathcal{B}, d\mu)$ we can define a measure on the measure space $(X, \mathcal{A}, \mu)$ by $\mu_{\mathcal{A}}(A) = \int_A f d\mu$, for $A \in \mathcal{A}$. Clearly, $\mu_{\mathcal{A}} << \mu$ (where μ is here defined on $\mathcal{A}$).

DEFINITION. By the Radon-Nikodym theorem there is a unique function $E(f|\mathcal{A}) := \frac{d\mu_{\mathcal{A}}}{d\mu} \in L^1(X, \mathcal{A}, d\mu)$ which is called the *conditional expectation*.

Since in general $\mathcal{A}$ is *strictly* contained in $\mathcal{B}$ then $E(f|\mathcal{A})$ may be very different from f, since it must be measurable on a smaller sigma-algebra. For example, if $\mathcal{A} = \{X, \emptyset\}$ then $E(f|\mathcal{A})$ is the constant function $\int f d\mu$.

The main properties of $E(f|\mathcal{A})$ are

(i) $\int_A E(f|\mathcal{A}) d\mu = \int_A f d\mu$ for all $A \in \mathcal{A}$,
(ii) $E(f|\mathcal{A}) \in L^1(X, \mathcal{A}, d\mu)$,

(The first two properties are just the definition repeated.),

(iii) if $f \in L^1(X, \mathcal{B}, \mu)$ and $g \in L^\infty(X, \mathcal{A}, d\mu)$ then $E(fg|\mathcal{A}) = gE(f|\mathcal{A})$,

(iv) if $f \in L^1(X, \mathcal{B}, \mu)$ and $\mathcal{A}_2 \subset \mathcal{A}_1 \subset \mathcal{B}$ then $E\left(E(f|\mathcal{A}_1)|\mathcal{A}_2\right) = E(f|\mathcal{A}_2)$,

(v) if $f \in L^1(X, \mathcal{B}, \mu)$ then $|E(f|\mathcal{A})| \leq E(|f||\mathcal{A})$, and if $f, g \in L^2(X, \mathcal{B}, d\mu)$ and $\frac{1}{p} + \frac{1}{q} = 1$ then $E(|fg||\mathcal{A}) \leq E(|f|^p|\mathcal{A})^{1/p} E(|g|^q|\mathcal{A})^{1/q}$,

(vi) If T preserves μ then $E(f|\mathcal{A})T = E(f \circ T|T^{-1}\mathcal{A})$, where $T^{-1}\mathcal{A} = \{T^{-1}A : A \in \mathcal{A}\}$.

Parts (iii)-(vi) are a trivial exercise from the definitions (cf. [5, p. 10]).

DEFINITION. Given any sub-sigma-algebra $\mathcal{A} \subset \mathcal{B}$ we can define the *conditional information function* $I(\alpha|\mathcal{A}) : X \to \mathbb{R}$ by

$$I(\alpha|\mathcal{A})(x) = -\sum_i \log \mu(A_i|\mathcal{A})(x)\chi_{A_i}(x)$$

where we write $\mu(A_i|\mathcal{A})(x) = E(\chi_{A_i}|\mathcal{A})(x)$ (called the *the conditional measure*).

Assume that we "know" the position of the point x relative to $\mathcal{A}$, then $I(\alpha|\mathcal{A})$ is an indicator of how much additional information we get from "knowing" the position of the point x relative to the partition α.

The following properties all come directly from the definition.

LEMMA 8.1.

(1) When $\mathcal{A} = \{\emptyset, X\}$ then $E(A_i|\{\emptyset, X\})(x) = \mu(A_i)$ and $I(\alpha|\mathcal{B})(x) = I(\alpha)(x)$.

(2) If $T : X \to X$ preserves the measure μ then $I(\alpha|\{\emptyset, X\})(Tx) = I(T^{-1}\alpha|T^{-1}\mathcal{A})(x)$ (where $T^{-1}\alpha = \{T^{-1}A_i\}$).

(3) If $\alpha \subset \mathcal{A}$ then $I(\alpha|\mathcal{A}) = 0$ almost everywhere.

We can associate to each partition α the sigma-algebra $\hat{\alpha}$ generated by α (in particular, $\hat{\alpha}$ is countable and consists of all unions of elements from α).

DEFINITION. Given two partitions α, β we define their *refinement*

$$\alpha \vee \beta = \{A_i \cap B_i : A_i \in \alpha, B_j \in \beta\}.$$

Given two sigma-algebras $\mathcal{A}, \mathcal{B}$ we denote by $\mathcal{A} \vee \mathcal{B}$ the sigma-algebra generated by $\{A \cap B : A \in \alpha, B \in \beta\}$.

The following lemma will prove very useful throughout this chapter.

LEMMA 8.2 (BASIC IDENTITY FOR INFORMATION). *Given partitions α, β and a third γ with associated sigma-algebra $\hat{\gamma}$ we have that*

$$I(\alpha \vee \beta | \hat{\gamma})(x) = I(\alpha | \hat{\beta} \vee \hat{\gamma})(x) + I(\beta | \hat{\gamma})(x)$$

(almost everywhere).

PROOF. Observe that for any function $g \in L^1(X, \mathcal{B}, \mu)$ we have that

$$E(g | \hat{\gamma})(x) = \sum_{C \in \gamma} \chi_C \frac{\int_C g(x) d\mu}{\mu(C)}.$$

In particular, for $B \in \beta$ we can set $g(x) = \chi_B(x)$ and then we get

$$\mu(B | \hat{\gamma})(x) = \sum_{C \in \gamma} \chi_C(x) \frac{\mu(B \cap C)}{\mu(C)}$$

and therefore

$$I(\beta | \hat{\gamma})(x) = - \sum_{C \in \gamma, B \in \beta} \chi_{C \cap B}(x) \log \left(\frac{\mu(B \cap C)}{\mu(C)} \right). \tag{8.1}$$

The partition $\beta \vee \gamma$ (with elements of the form $B \cap C$ with $B \in \beta, C \in \gamma$) gives that

$$I(\alpha | \hat{\gamma} \vee \hat{\beta})(x) = - \sum_{C \in \gamma, B \in \beta, A \in \alpha} \chi_{A \cap B \cap C}(x) \log \left(\frac{\mu(A \cap B \cap C)}{\mu(B \cap C)} \right). \tag{8.2}$$

Adding (8.1) and (8.2) gives that

$$I(\beta | \hat{\gamma})(x) + I(\alpha | \hat{\gamma} \vee \hat{\beta})(x)$$

$$= - \sum_{C \in \gamma, B \in \beta, A \in \alpha} \chi_{A \cap B \cap C}(x) \left(\log \left(\frac{\mu(B \cap C)}{\mu(C)} \right) + \log \left(\frac{\mu(A \cap B \cap C)}{\mu(B \cap C)} \right) \right)$$

$$= - \sum_{C \in \gamma, B \in \beta, A \in \alpha} \chi_{A \cap B \cap C}(x) \log \left(\frac{\mu(A \cap B \cap C)}{\mu(C)} \right)$$

$$= I(\alpha \vee \beta | \gamma)(x).$$

This completes the proof. ∎

If α and β are partitions we write $\alpha < \beta$ if every element of α is a union of elements of β. In this case $\alpha \vee \beta = \beta$.

COROLLARY 8.2.1. *If $\alpha < \beta$ then $I(\alpha | \hat{\gamma})(x) \leq I(\beta | \hat{\gamma})(x)$.*

8.2 The entropy of a partition

DEFINITION. We define the *entropy* of the partition α by

$$H(\alpha) = \int I(\alpha)(x)d\mu(x) = -\sum_{A\in\alpha} \mu(A)\log\mu(A).$$

Given a partition α and a sub-sigma-algebra $\mathcal{A} \subset \mathcal{B}$ we define the *conditional entropy* by $H(\alpha|\mathcal{A}) = \int I(\alpha|\mathcal{A})(x)d\mu(x)$.

LEMMA 8.3.
(1) When $\mathcal{A} = \{\emptyset, X\}$ then $H(\alpha|\{\emptyset, X\}) = H(\alpha)$.
(2) If $T : X \to X$ preserves the measure μ then $H(\alpha|\mathcal{A}) = H(T^{-1}\alpha|T^{-1}\mathcal{A})$.
(3) If $\alpha \subset \mathcal{A}$ then $H(\alpha|\mathcal{A}) = 0$.
(4) Given α, γ we have $H(\alpha|\hat{\gamma}) \leq H(\alpha)$.

PROOF. Parts (1), (2) and (3) follow by integrating the corresponding results for the information function in Lemma 8.1.
For part (4) we have

$$\begin{aligned}
H(\alpha|\gamma) &= -\sum_{A\in\alpha, C\in\gamma} \mu(A\cap C)\log\left(\frac{\mu(A\cap C)}{\mu(C)}\right) \\
&\leq -\sum_{A\in\alpha, C\in\gamma} \mu(A\cap C)\log\left(\mu(A\cap C)\right) \\
&\leq -\sum_{A\in\alpha} \mu(A)\log\mu(A) = H(\alpha)
\end{aligned}$$

since for fixed $A \in \alpha$ we can bound $-\sum_{C\in\gamma} \mu(A\cap C)\log\left(\mu(A\cap C)\right) \leq -\mu(A)\log\mu(A)$ using concavity of $t \mapsto -t\log t$. (A more general result appears in Lemma 8.13.) ∎

LEMMA 8.4 (BASIC IDENTITY FOR ENTROPY). *Given partitions α, β and a third γ with associated sigma-algebra $\hat{\gamma}$ we have that*

$$H(\alpha \vee \beta|\hat{\gamma}) = H(\alpha|\hat{\beta} \vee \hat{\gamma}) + H(\beta|\hat{\gamma}).$$

COROLLARY 8.4.1 ("MONOTONICITY" OF ENTROPY FOR PARTITIONS). *Given two partitions α, β with $\alpha < \beta$ we have that $H(\alpha|\hat{\gamma}) \leq H(\beta|\hat{\gamma})$ (and, in particular $H(\alpha) \leq H(\beta)$)*

With the next definition, we begin to re-introduce measure preserving transformations.

DEFINITION. Assume that $T : X \to X$ preserves μ. Given a partition $\alpha = \{A_i\}$ we write

$$\vee_{i=0}^{n-1} T^{-i}\alpha = \{A_{r_0} \cap T^{-1}A_{r_1} \cap \cdots \cap T^{-(n-1)}A_{r_{n-1}} : A_{r_i} \in \alpha, i = 0, \ldots, n-1\}.$$

NOTATIONAL COMMENT. Frequently it proves convenient to drop the circumflex (hat) over $\hat{\gamma}$. Thus if we write $H(\alpha|\beta)$, say, we understand this to mean $H(\alpha|\hat{\beta})$.

For $n \geq 1$ we can write $H_n(\alpha) = H(\vee_{i=0}^{n-1}T^{-i}\alpha)$. By the above estimates we have that

$$
\begin{aligned}
H_{n+m}(\alpha) &= H\left(\vee_{i=0}^{n+m-1}T^{-i}\alpha\right) \\
&= H\left(\vee_{i=0}^{n-1}T^{-i}\alpha\right) + H\left(\vee_{i=n}^{n+m-1}T^{-i}\alpha \,\big|\, \vee_{i=0}^{n-1} T^{-i}\alpha\right) \\
&\leq H\left(\vee_{i=0}^{n-1}T^{-i}\alpha\right) + H\left(\vee_{i=n}^{n+m-1}T^{-i}\alpha\right) \\
&= H_n(\alpha) + H_m(\alpha).
\end{aligned}
\tag{8.3}
$$

Thus the sequence $H_n(\alpha)$, $n \geq 1$, is subadditive (which shows that the limit in the following definition exists).

DEFINITION. We define the *entropy of the partition* α relative to the transformation $T : X \to X$ as the limit $h(T,\alpha) = \lim_{n \to +\infty} \frac{H_n(\alpha)}{n}$.

Notice that in particular from (8.3) we have that $0 \leq h(T,\alpha) \leq H(\alpha)$. The following result gives an equivalent characterization.

PROPOSITION 8.5 (ALTERNATIVE DEFINITION OF $h(T,\alpha)$).

$$h(T,\alpha) = \lim_{n \to +\infty} H(\alpha \,|\, \vee_{i=1}^{n-1} T^{-i}\alpha).$$

(N.B. Sometimes it is convenient to write this limit as $H(\alpha \,|\, \vee_{i=1}^{\infty} T^{-i}\alpha)$.)

PROOF. Using Lemma 8.4 we see that

$$
\begin{aligned}
H(\vee_{i=0}^{n-1}T^{-i}\alpha) &= H(\alpha \,|\, \vee_{i=1}^{n-1} T^{-i}\alpha) + H(\vee_{i=0}^{n-2}T^{-i}\alpha) \\
&= H(\alpha \,|\, \vee_{i=1}^{n-1} T^{-i}\alpha) + H(\alpha \,|\, \vee_{i=1}^{n-2} T^{-i}\alpha) + H(\vee_{i=0}^{n-3}T^{-i}\alpha) \\
&\quad \cdots \\
&= \sum_{r=1}^{n} H(\alpha \,|\, \vee_{i=1}^{r-1} T^{-i}\alpha) + H(\alpha).
\end{aligned}
$$

We then see that

$$\lim_{n \to +\infty} \frac{1}{n} H(\vee_{i=0}^{n-1} T^{-i} \alpha) = \lim_{n \to +\infty} H(\alpha | \vee_{i=1}^{n-1} T^{-i} \alpha)$$

as required (since if $a_n \to a$ for any sequence of real numbers then $\frac{a_1 + \ldots + a_n}{n} \to a$).

∎

The entropy of the transformation relative to two different partitions is described by the following inequality.

LEMMA 8.6. *For finite entropy partitions α, β of X we have that*

$$h(T, \alpha) \leq h(T, \beta) + H(\alpha | \beta).$$

PROOF. Since $\left(\vee_{i=0}^{n-1} T^{-i} \alpha\right) \vee \left(\vee_{i=0}^{n-1} T^{-i} \beta\right) > \vee_{i=0}^{n-1} T^{-i} \alpha$ we have that

$$H\left(\vee_{i=0}^{n-1} T^{-i} \alpha\right) \leq H\left(\left(\vee_{i=0}^{n-1} T^{-i} \alpha\right) \vee \left(\vee_{i=0}^{n-1} T^{-i} \beta\right)\right)$$
$$= H\left(\vee_{i=0}^{n-1} T^{-i} \beta\right) + H\left(\vee_{i=0}^{n-1} T^{-i} \alpha | \vee_{i=0}^{n-1} T^{-i} \beta\right)$$

(where we use Lemma 8.4, with γ being the trivial partition, for the last line). We next estimate

$$H\left(\vee_{i=0}^{n-1} T^{-i} \alpha | \vee_{i=0}^{n-1} T^{-i} \beta\right)$$
$$= H\left(\alpha | \vee_{i=0}^{n-1} T^{-i} \beta\right) + H\left(\vee_{i=1}^{n-1} T^{-i} \alpha | \alpha \vee \left(\vee_{i=0}^{n-1} T^{-i} \beta\right)\right)$$
$$\leq H(\alpha | \beta) + H\left(\vee_{i=1}^{n-1} T^{-i} \alpha | \vee_{i=1}^{n-1} T^{-i} \beta\right)$$
$$\leq H(\alpha | \beta) + H\left(\vee_{i=0}^{n-2} T^{-i} \alpha | \vee_{i=0}^{n-2} T^{-i} \beta\right).$$

Proceeding inductively gives us

$$H\left(\vee_{i=0}^{n-1} T^{-i} \alpha | \vee_{i=0}^{n-1} T^{-i} \beta\right) \leq n H(\alpha | \beta).$$

Finally, we see that

$$\frac{1}{n} H\left(\vee_{i=0}^{n-1} T^{-i} \alpha\right) \leq \frac{1}{n} H\left(\vee_{i=0}^{n-1} T^{-i} \beta\right) + H(\alpha | \beta).$$

Letting $n \to +\infty$ gives the correct inequality.

∎

COROLLARY 8.6.1. *For finite entropy partitions α, β of X we have that*

$$|h(T, \beta) - h(T, \alpha)| \leq H(\beta | \alpha) + H(\alpha | \beta).$$

PROOF. By interchanging α and β in Lemma 8.6 we get that $h(T, \beta) \leq h(T, \alpha) + H(\beta | \alpha)$.

8.3 The entropy of a transformation

Consider a measure preserving transformation $T : X \to X$ on a probability space $(X, \mathcal{B}, \mu)$. We want to associate to this a numerical invariant. We start from the definition of the entropy relative to a partition α and then remove the dependence on α by taking a supremum.

DEFINITION. We define the *measure theoretic entropy* of $T : X \to X$ for the probability space $(X, \mathcal{B}, \mu)$ by $h_\mu(T) = \sup_{\{\alpha : H(\alpha) < +\infty\}} h(T, \alpha)$.

We write the measure μ as a subscript not only to remind us that there is an ambient measure, but also to distinguish the notation from that of topological entropy in chapter 3.

As one might imagine, it can be very difficult to compute the measure theoretic entropy from the definition given. We now want to describe a very important method of *practical* computation. We begin with a result which replaces the supremum in the definition of the measure theoretic entropy with a limit.

LEMMA 8.7 (ABRAMOV). *Let $\beta_1 \subset \beta_2 \subset \ldots \subset \beta_k \subset \mathcal{B}$ be an increasing sequence of partitions with $H(\beta_k) < +\infty$, $\forall k \geq 1$; and such that $\cup_n \beta_k$ generates the sigma-algebra $\mathcal{B}$. Then $h_\mu(T) = \lim_{k \to +\infty} h(T, \beta_k)$.*

We shall return to the proof in section 8.7.

The following definition gives us a way to generate the increasing partitions.

DEFINITION. We say that a partition α with $H(\alpha) < +\infty$ is called a *strong generator* for the probability space $(X, \mathcal{B}, \mu)$ if $\vee_{i=0}^{\infty} T^{-i}\alpha = \mathcal{B}$.

If T is invertible, then we say that a partition α with $H(\alpha) < +\infty$ is called a *generator* for the probability space $(X, \mathcal{B}, \mu)$ if $\vee_{i=-\infty}^{\infty} T^{-i}\alpha = \mathcal{B}$

LEMMA 8.8 (SINAI). *If α is a (strong) generator then $h_\mu(T) = h(T, \alpha)$.*

We shall return to the proof in section 8.7. Before developing the theory needed to prove these two results we shall use them to compute the measure theoretic entropy of some simple examples.

Example 1 (doubling map). Let $X = \mathbb{R}/\mathbb{Z}$, $\mathcal{B}$ denote the Borel sigma-algebra, and μ the Haar-Lebesgue measure. We let $T : X \to X$ be the doubling map $T(x) = 2x \pmod 1$. Let $\alpha = \{[0, \frac{1}{2}), [\frac{1}{2}, 1)\}$; then observe that

$$\alpha \vee T^{-1}\alpha = \left\{ \left[0, \frac{1}{4}\right), \left[\frac{1}{4}, \frac{1}{2}\right), \left[\frac{1}{2}, \frac{3}{4}\right), \left[\frac{3}{4}, 1\right) \right\}$$

and more generally,

$$\vee_{i=0}^{n-1} T^{-i}\alpha = \left\{ \left[\frac{i}{2^n}, \frac{i+1}{2^n} \right) : i = 0, \dots, 2^n - 1 \right\}.$$

We can now calculate

$$
\begin{aligned}
H(\vee_{i=0}^{n-1} T^{-i}\alpha) &= - \sum_{i=0}^{2^n-1} \mu \left(\left[\frac{i}{2^n}, \frac{i+1}{2^n} \right) \right) \log \left(\left[\frac{i}{2^n}, \frac{i+1}{2^n} \right) \right) \\
&= - \sum_{i=0}^{2^n-1} \left(\frac{1}{2^n} \right) \log \left(\frac{1}{2^n} \right) \\
&= - 2^n \left(\frac{1}{2^n} \right) \log \left(\frac{1}{2^n} \right) \\
&= n \log 2.
\end{aligned}
$$

Thus we see that $\frac{1}{n} H(\vee_{i=0}^{n-1} T^{-i}\alpha) = \log 2$ and thus letting $n \to +\infty$ gives that $h_\mu(T) = \log 2$.

Example 2 (rotations on the circle). Let $X = \mathbb{R}/\mathbb{Z}$, let $\mathcal{B}$ denote the Borel sigma-algebra, and let μ be the Haar-Lebesgue measure. We let $T : X \to X$ be the rotation $T(x) = x + a \pmod 1$ for some fixed values $a \in \mathbb{R}$.

First assume that $a = \frac{p}{q}$ is a rational number. For any partition β we see that $T^{-q}\beta = \beta$. Thus

$$\vee_{k=0}^{nq-1} T^{-k}\beta = \vee_{k=0}^{q-1} T^{-k}\beta$$

and so in particular

$$
\begin{aligned}
h_\mu(T, \beta) &= \lim_{n \to +\infty} \frac{1}{qn} H \left(\vee_{k=0}^{nq-1} T^{-k}\beta \right) \\
&= \lim_{n \to +\infty} \frac{1}{qn} H \left(\vee_{k=0}^{q-1} T^{-k}\beta \right) \\
&= 0.
\end{aligned}
$$

Thus the measure theoretic entropy of any partition is zero, and thus the measure theoretic entropy of the transformation, $h_\mu(T) = \sup_{H(\beta)<+\infty} h(T, \beta) = 0$.

Next, assume that a is irrational. We let $\beta = \{[0, \frac{1}{2}), [\frac{1}{2}, 1)\}$. Since the sequence $\frac{1}{2} + na \pmod 1$ is dense in the unit circle (Weyl's theorem) we see that the partition is (strong) generating. Moreover, we see that $\mathcal{B} = \vee_{k=1}^\infty T^{-k}\beta$. As we observed before

$$h(T, \beta) = \lim_{n \to +\infty} H(\beta | \vee_{k=1}^{n-1} T^{-k}\beta) = 0$$

and therefore $h_\mu(T) = h(T, \beta) = 0$.

As one might imagine, a similar method applies to rotation on tori $\mathbb{R}^n / \mathbb{Z}^n$, $n \geq 2$.

Example 3 (Markov measures). Let $\sigma : X \to X$ denote a subshift of finite type

$$X = \{(x_n) \in \prod_{-\infty}^{+\infty} \{1, \dots, k\} : A_{x_n x_{n+1}} = 1\}$$

where $A = (A_{ij})$ is a $k \times k$ matrix with entries either zero or unity.

We associate to this a $k \times k$ stochastic matrix $P = (P_{ij})$ (cf. Example 4 in section 7.3) and let $p = (p_1, \dots, p_k)$ be the left eigenvector associated to the left eigenvalue unity.

The partition $\alpha = \{[1], \dots [k]\}$ for X is generating. Let $\sigma : X \to X$ denote the shift transformation. The refined partition $\vee_{k=0}^{n-1} T^{-k} \alpha$ consists of "cylinder" sets of the form

$$[z_0, \dots, z_{n-1}] = \{(x_n)_{n \in \mathbb{Z}} \in X : x_i = z_i, 0 \leq i \leq n - 1\}$$

where $z_i \in \{1, \dots, k\}$.

By the definition of the Markov measure μ associated to P we have that

$$\mu([z_0, \dots, z_{n-1}]) = p_{z_0} P_{z_0 z_1} P_{z_1 z_2} \dots P_{z_{n-2} z_{n-1}}.$$

We explicitly compute:

$$H(\vee_{k=0}^{n-1} T^{-k} \alpha)$$

$$= - \sum_{[z_0, \dots, z_{n-1}]} \mu([z_0, \dots, z_{n-1}]) \log \mu([z_0, \dots, z_{n-1}])$$

$$= - \sum_{[z_0, \dots, z_{n-1}]} p_{z_0} P_{z_0 z_1} P_{z_1 z_2} \dots P_{z_{n-2} z_{n-1}} \log \left(p_{z_0} P_{z_0 z_1} P_{z_1 z_2} \dots P_{z_{n-2} z_{n-1}} \right)$$

$$= - \sum_{[z_0, \dots, z_{n-1}]} p_{z_0} P_{z_0 z_1} P_{z_1 z_2} \dots P_{z_{n-2} z_{n-1}} (\log p_{z_0} + \log P_{z_0 z_1} + \log P_{z_1 z_2}$$

$$+ \log P_{z_2 z_3} + \dots + \log P_{z_{n-2} z_{n-1}})$$

$$= - \sum_{i=1}^{k} p_i \log p_i - (n - 1) \sum_{i,j=1}^{k} p_i P_{ij} \log P_{ij}$$

(where we use that $pP = p$ and P is stochastic).

Therefore we see that

$$h_\mu(T) = \lim_{n \to +\infty} \frac{1}{n} H(\vee_{k=0}^{n-1} T^{-k} \alpha) = - \sum_{i,j=1}^{k} p_i P_{ij} \log P_{ij}.$$

In the special case that $X = X_k = \prod_{-\infty}^{+\infty}\{1,\dots,k\}$ we can define a "Bernoulli measure" from a probability vector $(p_1,\dots,p_k)$ (i.e. $(p_1 + \dots + p_k = 1)$ by

$$\mu([z_0,\dots,z_{n-1}]) = p_{z_0}p_{z_1}p_{z_2}\cdots p_{z_{n-1}}.$$

The measure theoretic entropy in this case is $h_\mu(T) = \sum_{i=1}^{k} p_i \log p_i$. For example, where $X = X_2$ and $p = (\frac{1}{2},\frac{1}{2})$ we have that $h_\mu(T) = \log 2$. Where $X = X_3$ and $p = (\frac{1}{3},\frac{1}{3},\frac{1}{3})$ we have that $h_\mu(T) = \log 3$, etc.

8.4 The increasing martingale theorem

We now begin to develop some of the machinery need to prove the results in section 8.3.

We know that if $f \in L^1(X,\mathcal{B},\mu)$ and $\mathcal{A} \subset \mathcal{B}$ is a sub-sigma-algebra then we can associate the conditional expectation $E(f|\mathcal{A}) \in L^1(X,\mathcal{A},\mu)$. The increasing martingale theorem describes how $E(f|\mathcal{A})$ depends on the sigma-algebra $\mathcal{A}$. This is crucial in understanding the corresponding behaviour of the information function and thus the measure theoretic entropy.

The following simple lemma is very useful.

LEMMA 8.9. *If $(X,\mathcal{B},\mu)$ is a probability space and if $\mathcal{B}_1 \subset \mathcal{B}_2 \subset \dots \subset \mathcal{B}_N \subset \mathcal{B}$ are sigma-algebras and $\lambda > 0$ then if we let*

$$E = \{x \in X \colon \max_{1 \le n \le N} E(f|\mathcal{B}_n)(x) > \lambda\}$$

then we have the upper bound on its measure $\mu(E) \le \frac{1}{\lambda}\int|f|d\mu$ with $f \in L^1(X,\mathcal{B},\mu)$.

PROOF. Without loss of generality we can assume that $f \ge 0$ (otherwise we replace f by $\max\{f(x),0\}$). We can partition $E = E_1 \cup \dots \cup E_N$ where

$$E_n = \{x \in X : E(f|\mathcal{B}_n)(x) > \lambda, E(f|\mathcal{B}_i)(x) \le \lambda, i = 1,2,\dots,n-1\}$$

(and observe that $E_n \in \mathcal{B}_n$); then $E_i \cap E_j = \emptyset$ for $i \ne j$. We then write

$$\int_E f d\mu = \sum_{n=1}^{N}\int_{E_n} f d\mu = \sum_{n=1}^{N}\int_{E_n} E(f|\mathcal{B}_n)d\mu \ge \sum_{n=1}^{N}\lambda\mu(E_n) = \lambda\mu(E).$$

Thus $\mu(E) \le \frac{1}{\lambda}\int f d\mu = \frac{1}{\lambda}\int|f|d\mu$. ∎

REMARK. This is very similar to the Chebyshev inequality for $f \in L^1(X,\mathcal{B},\mu)$ and $\lambda > 0$ which says that $\mu\{x \in X : f(x) > \lambda\} \le \frac{\int|f|d\mu}{\lambda}$.

This brings us to the main result of this section.

THEOREM 8.10 (INCREASING MARTINGALE THEOREM). *Let $f \in L^1(X,$ $\mathcal{B}, \mu)$. Assume that $\mathcal{B}_1 \subset \mathcal{B}_2 \subset \ldots \subset \mathcal{B}_n \subset \ldots \subset \mathcal{B}$ is an increasing sequence of sigma-algebras and that the union $\cup_{n=1}^{\infty} \mathcal{B}_n$ generates $\mathcal{B}$ (written $\mathcal{B}_n \rightarrow \mathcal{B}$). Then $E(f|\mathcal{B}_n) \rightarrow f$ in $L^1(X, \mathcal{B}, \mu)$ and $E(f|\mathcal{B}_n)(x) \rightarrow f(x)$, almost everywhere.*

PROOF. The theorem is clearly true on the subspace $\cup_{k=1}^{\infty} L^1(X, \mathcal{B}_k, \mu)$ since if $g \in L^1(X, \mathcal{B}_k, \mu)$ then $E(g|\mathcal{B}_n) = g$ for $n \geq k$. Moreover, this subspace is dense in $L^1(X, \mathcal{B}, \mu)$ in the L^1 norm.

Given an arbitrary $f \in L^1(X, \mathcal{B}, \mu)$ we can choose $\epsilon > 0$ and $g \in L^1(X, \mathcal{B}_k, \mu)$, say, with $\int |f - g| d\mu < \epsilon$. We then see that for any $n \geq k$ we have that

$$\int |E(f|\mathcal{B}_n) - f| d\mu$$

$$\leq \int |E(f|\mathcal{B}_n) - E(g|\mathcal{B}_n)| d\mu + \int |E(g|\mathcal{B}_n) - g| d\mu + \int |g - f| d\mu$$

$$\leq 2 \int |g - f| d\mu$$

(where $\int |E(g|\mathcal{B}_n) - g| d\mu = 0$ since $E(g|\mathcal{B}_n) = g$ and we use that $E(.|\mathcal{B}_n)$ is a contraction on $L^1(X, \mathcal{B}, \mu)$). In particular, $\limsup_{n \to +\infty} \int |E(f|\mathcal{B}_n) - f| d\mu \leq 2\epsilon$. Since $\epsilon > 0$ is arbitrary, we see that we have L^1 convergence.

To show that we also have almost everywhere convergence, we argue as follows:

$$\mu\{x \in X : \limsup_{n \to +\infty} |E(f|\mathcal{B}_n)(x) - f(x)| > \epsilon^{1/2}\}$$

$$\leq \mu\{x \in X : \limsup_{n \to +\infty} (|E((f - g)|\mathcal{B}_n)(x) - (f - g)(x)|$$

$$+ |E(g|\mathcal{B}_n)(x) - g(x)|) > \epsilon^{1/2}\}$$

$$\leq \mu\{x \in X : \limsup_{n \to +\infty} |E((f - g)|\mathcal{B}_n)(x)| + |(f - g)(x)| > \epsilon^{1/2}\}$$

$$\leq \mu\{x \in X : \limsup_{n \to +\infty} |E((f - g)|\mathcal{B}_n)(x)| > \frac{1}{2}\epsilon^{1/2}\}$$

$$+ \mu\{x \in X : |(f - g)(x)| > \frac{1}{2}\epsilon^{1/2}\}$$

$$\leq 2 \left(\frac{1}{\frac{1}{2}\epsilon^{1/2}} \right) \int |f - g| d\mu \leq 2 \left(\frac{1}{\frac{1}{2}\epsilon^{\frac{1}{2}}} \right) \epsilon \leq 4\epsilon^{1/2}$$

(where we have used Lemma 8.9 and the Chebyshev inequality). Since $\epsilon > 0$ is arbitrary this shows almost everywhere convergence. ∎

REMARK. There is a corresponding "decreasing martingale theorem", but we shall not need it.

8.5 Entropy and sigma-algebras

We want to apply the increasing martingale theorem to the information functions. First we need a simple technical lemma.

LEMMA 8.11. *If α is a partition with $H(\alpha) < +\infty$ and we have sub-sigma-algebras $\mathcal{A}_1 \subset \mathcal{A}_2 \subset \ldots \subset \mathcal{B}$ then*

$$\int \left(\sup_{n \geq 1} I(\alpha|\mathcal{A}_n) \right) d\mu \leq H(\alpha) + 1$$

(and, in particular, $f(x) = \sup_{n \geq 1} I(\alpha|\mathcal{A}_n)(x) \in L^1(X, \mathcal{B}, \mu)$).

PROOF. We can write

$$\int f(x) d\mu(x) = \int_0^\infty F(t) dt \tag{8.4}$$

where $F(t) = \mu \{ x \in X : f(x) > t \}$, provided the right hand side of (8.1) is finite.

We can write

$$F(t) = \mu \{ x \in X : \sup_{n \geq 1} I(\alpha|\mathcal{A}_n)(x) > t \}$$

$$= \mu \left\{ x \in X : \sup_{n \geq 1} \left(- \sum_{A \in \alpha} \chi_A(x) \log \mu(A|\mathcal{A}_n)(x) \right) > t \right\}$$

$$= \sum_{A \in \alpha} \mu \left(A \cap \{ x \in X : \sup_{n \geq 1} (- \log \mu(A|\mathcal{A}_n)(x)) > t \} \right)$$

(since the sets A are disjoint). However, we can simplify this by writing

$$\{ x \in X : \sup_{n \geq 1} (- \log \mu(A|\mathcal{A}_n)(x)) > t \}$$

$$= \{ x \in X : \inf_{n \geq 1} (\log \mu(A|\mathcal{A}_n)(x)) < -t \}$$

$$= \{ x \in X : \inf_{n \geq 1} (\mu(A|\mathcal{A}_n)(x)) < e^{-t} \} = \cup_{n \geq 1} A_n$$

where $A_n = \{ x \in X : \mu(A|\mathcal{A}_n)(x) < e^{-t}$ and $\mu(A|\mathcal{A}_i)(x) \geq e^{-t}$, for $i = 1, \ldots, n-1 \}$ are disjoint sets. If we write

$$F(t) = \sum_{A \in \alpha} \mu (A \cap (\cup_{n \geq 1} A_n)) = \sum_{A \in \alpha} \sum_{n \geq 1} \mu (A \cap A_n)$$

then we can use the estimates

$$\mu(A \cap A_n) = \int_{A_n} \chi_A d\mu = \int_{A_n} E(\chi_A|\mathcal{A}_n) d\mu \leq \int_{A_n} e^{-t} d\mu = e^{-t} \mu(A_n).$$

We now have two possible upper bounds on the same summation:

$$\sum_{n \geq 1} \mu(A \cap A_n) \leq \sum_n e^{-t} \mu(A_n) = e^{-t} \quad \text{and} \quad \sum_{n \geq 1} \mu(A \cap A_n) \leq \mu(A).$$

Therefore $F(t) \leq \sum_{A \in \alpha} \min\{e^{-t}, \mu(A)\}$. Finally, we can use this bound to estimate

$$\int_0^\infty F(t)dt \leq \int_0^\infty \left(\sum_{A \in \alpha} \min\{e^{-t}, \mu(A)\} \right) dt$$

$$= -\sum_{A \in \alpha} \left(\mu(A) \log \mu(A) - \int_{-\log \mu(A)}^\infty e^{-t}dt \right)$$

$$= -\sum_{A \in \alpha} (\mu(A) \log \mu(A) - \mu(A))$$

$$= H(\alpha) + 1.$$

■

We are now in a position to prove the following crucial result.

THEOREM 8.12. *If α is a partition with $H(\alpha) < +\infty$ and $\mathcal{A}_1 \subset \mathcal{A}_2 \subset \ldots \to \mathcal{B}$ is an increasing sequence of sub-sigma-algebra then $I(\alpha|\mathcal{A}_n)(x) \to I(\alpha|\mathcal{B})(x)$ almost everywhere and in L^1. Thus $H(\alpha|\mathcal{A}_n) \to H(\alpha|\mathcal{B})$ as $n \to +\infty$.*

PROOF. By Theorem 8.10 $\mu(A|\mathcal{A}_n) \to \mu(A|\mathcal{B})$ almost everywhere, for any $A \in \alpha$. This implies that $I(\alpha|\mathcal{A}_n)(x) \to I(\alpha|\mathcal{B})(x)$ almost everywhere.

By Lemma 8.11 we have that $I(\alpha|\mathcal{A}_n)$ are dominated by the integrable function $\sup_{n \geq 1} I(\alpha|\mathcal{A}_n)(x)$. Thus by Lebesgue's dominated convergence theorem we have that $I(\alpha|\mathcal{A}_n)(x) \to I(\alpha|\mathcal{B})(x)$ in L^1 (i.e. $\int |I(\alpha|\mathcal{A}_n)(x) - I(\alpha|\mathcal{B})(x)|d\mu(x) \to 0$ as $n \to +\infty$).

Integrating shows the corresponding result for measure theoretic entropy i.e. $H(\alpha|\mathcal{A}_n)(x) \to H(\alpha|\mathcal{B})(x)$ as $n \to +\infty$.

■

Using Theorem 8.12 we can extend the basic identities (Lemma 8.2 and Lemma 8.4) to arbitrary sub-sigma-algebras $\mathcal{C} \subset \mathcal{B}$, i.e.

(1) for the information functions

$$I(\alpha \vee \beta|\mathcal{C}) = I(\alpha|\hat{\beta} \vee \mathcal{C}) + I(\beta|\mathcal{C})$$

(and, in particular, $I(\alpha|\mathcal{C}) \geq I(\beta|\mathcal{C})$ if $\alpha > \beta$),

(2) for the measure theoretic entropy

$$H_\mu(\alpha \vee \beta|\mathcal{C}) = H_\mu(\alpha|\hat{\beta} \vee \mathcal{C}) + h_\mu(\beta|\mathcal{C})$$

(and, in particular, $H_\mu(\alpha|\mathcal{C}) \geq H_\mu(\beta|\mathcal{C})$ if $\alpha > \beta$).

This only requires that we choose partitions γ such that $\hat{\gamma} \to \mathcal{C}$ and apply the theorem (and this is a basic property of Lebesgue spaces).

8.6 Conditional entropy

We want to consider how changing the sigma-algebra (with the same partition) affects the conditional measure theoretic entropy. The following lemmas are useful.

LEMMA 8.13. *Assume that $f \in L^1(X, \mathcal{B}, \mu)$ and $0 \leq f(x) \leq 1$ a.e. and that $\mathcal{A} \subset \mathcal{B}$ is a sub-sigma-algebra. Let $\psi : [0, 1] \to \mathbb{R}$ be a concave function (i.e. $\psi(\alpha x + (1-\alpha)y) \geq \alpha\psi(x) + (1-\alpha)\psi(y)$); then $\psi(E(f|\mathcal{A})) \geq E(\psi(f)|\mathcal{A})$.*

PROOF. First consider the case of simple functions $f(x) = \sum_{i=1}^{n} b_i \chi_{B_i}$, where $\{B_1, \ldots, B_n\}$ is a partition for X.
By linearity of $E(.|\mathcal{A})$,

$$E(f|\mathcal{A})(x) = \sum_{i=1}^{n} b_i E(\chi_{B_i}|\mathcal{A})(x) = \sum_{i=1}^{n} b_i \mu(B_i|\mathcal{A})(x)$$

and observe that $\sum_{i=1}^{n} \mu(B_i|\mathcal{A})(x) = 1$. We can compute

$$\psi\left(E(f|\mathcal{A})\right) \geq \sum_{i=1}^{n} \psi(b_i)\mu(B_i|\mathcal{A}) = E\left(\sum_{i=1}^{n} \psi(b_i)\chi_{B_i}|\mathcal{A}\right) = E(\psi(f)|\mathcal{A}). \quad (8.5)$$

For an arbitrary function $f \in L^1(X, \mathcal{B}, \mu)$ we can choose a monotonically increasing sequence of step functions f_k increasing to f a.e. Since $E(\cdot|\mathcal{A})$ takes positive functions to positive functions we see that $E(f_k|\mathcal{A}) \to E(f|\mathcal{A})$. We can now take limits in (8.5) to get that $\psi\left(E(f|\mathcal{A})\right) \geq E(\psi(f)|\mathcal{A})$, as required. ∎

The following is a simple application of this result.

PROPOSITION 8.14. *If β is a partition and $\mathcal{A}_2 \subset \mathcal{A}_1 \subset \mathcal{B}$ are sub-sigma-algebras then $H(\beta|\mathcal{A}_1) \leq H(\beta|\mathcal{A}_2)$.*

(The corresponding result for information functions may not be true).

PROOF. For each $B \in \beta$ we fix the choice $f = \mu(B|\mathcal{A}_1)$. We then fix $\psi(t) = -t\log(t)$ for $0 < t \leq 1$ (and $\psi(0) = 0$). We then have that $\psi(f) = -\mu(B|\mathcal{A}_1)\log(\mu(B|\mathcal{A}_1))$ and the lemma (Jenson's inequality) gives that

$$-E\left(\mu(B|\mathcal{A}_1)\log(\mu(B|\mathcal{A}_1))|\mathcal{A}_2\right) \leq -\mu(B|\mathcal{A}_2)\log\mu(B|\mathcal{A}_2).$$

Integrating both sides with respect to μ (and summing over $B \in \beta$) gives that $H(\beta|\mathcal{A}_1) \leq H(\beta|\mathcal{A}_2)$. ∎

8.7 Proofs of Lemma 8.7 and Lemma 8.8

We can now use the results from the preceding sections to supply the omitted proofs of Lemmas 8.7 and 8.8.

PROOF OF LEMMA 8.7. We know that $h(T, \beta_n) \leq h(T, \beta_m) + H(\beta_n | \hat{\beta}_m)$ for $n, m \geq 1$. Moreover, if $m \geq n$ then $\beta_n \subset \beta_m$ and so $H(\beta_n | \beta_m) = 0$. Thus $h(T, \beta_n)$ is monotonically decreasing and so converges.

For any partition α with $H(\alpha) < +\infty$ we can start with the inequality $h(T, \alpha) \leq h_\mu(T, \beta_n) + H(\alpha | \beta_n)$ (proved in a previous lemma). By a corollary to the increasing martingale theorem we know that $H(\alpha | \beta_n) \to H(\alpha | \mathcal{B}) = 0$ (since $\alpha \subset \mathcal{B}$). We conclude that

$$h(T, \alpha) \leq \limsup_{n \to +\infty} h(T, \beta_n).$$

Taking the supremum over all such α gives that

$$h_\mu(T) = \sup_\alpha h(T, \alpha) \leq \lim_{n \to +\infty} h(T, \beta_n).$$

Clearly, $h_\mu(T) \geq h(T, \beta_n)$ for $n \geq 1$. Thus

$$h_\mu(T) \geq \sup_n h(T, \beta_n) \geq \lim_{n \to +\infty} h(T, \beta_n)$$

and the proof is complete.

∎

Before moving on to the proof of Lemma 8.8, we recall the following useful fact.

A FACT ABOUT LEBESGUE SPACES. For Lebesgue spaces a necessary and sufficient condition for $\beta_n \to \mathcal{B}$ is that there exists a set of zero measure $N \subset X$ such that for $x, y \in X - N$ (with $x \neq y$) there exist $n \geq 1$ and $B \in \beta_n$ such that $x \in B$ but $y \notin B$.

PROOF OF LEMMA 8.8. This is an application of Abramov's result where we take $\beta_n = \vee_{i=-n}^n T^{-i} \alpha$ or $\beta_n = \vee_{i=0}^n T^{-i} \alpha$, as appropriate. We then have that

$$\begin{aligned} h(T, \beta_k) &= H(\beta_k | \vee_{i=1}^\infty T^{-i} \beta_k) \\ &= H(\vee_{i=1}^{k-1} T^{-i} \alpha | \vee_{i=1}^\infty T^{-i} \alpha) \\ &= h(T, \alpha) \end{aligned}$$

and we need only let $k \to +\infty$.

∎

8.8 Isomorphism

Entropy is very important in the classification of measure preserving transformations. We begin with a definition.

DEFINITION. Let $(X_i, \mathcal{B}_i, \mu_i)$, for $i = 1, 2$, be probability spaces; then an *isomorphism* between measure preserving transformations $T_1 : X_1 \to X_1$ and $T_2 : X_2 \to X_2$ is a map $\phi : X_1 \to X_2$ such that

(1) ϕ is a bijection (after removing sets of zero measure, if necessary),
(2) both ϕ and ϕ^{-1} are measurable (i.e. $\phi^{-1}\mathcal{B}_2 \subset \mathcal{B}_1$ and $\phi\mathcal{B}_1 \subset \mathcal{B}_2$),
(3) $\mu_1(\phi^{-1}B) = \mu_2(B)$ for $B \in \mathcal{B}_2$ (also $\mu_2(\phi B) = \mu_1(B)$ for $B \in \mathcal{B}_1$),
(4) $\phi \circ T_1 = T_2 \circ \phi$.

THEOREM 8.15. *Entropy is an isomorphism invariant (i.e. if $T_1 : X_1 \to X_1$ and $T_2 : X_2 \to X_2$ are isomorphic then $h_{\mu_1}(T_1) = h_{\mu_2}(T_2)$).*

PROOF. Let α be a partition for X_2; then clearly $\phi^{-1}\alpha = \{\phi^{-1}(A) : A \in \alpha\}$ is a partition for X_1. Moreover, the properties for ϕ imply that

$$\frac{1}{n} H \left(\vee_{i=0}^{n-1} T_2^{-i}\alpha \right) = \frac{1}{n} H \left(\vee_{i=0}^{n-1} T_1^{-i}\phi^{-1}\alpha \right).$$

Letting $n \to +\infty$ gives that $h_{\mu_1}(T_1, \phi^{-1}\alpha) = h_{\mu_2}(T_2, \alpha)$. Taking the supremum over all partitions α (with $H(\alpha) < +\infty$) gives the result. ∎

EXAMPLE (BERNOULLI SHIFTS). Using the formula for the entropy of a Bernoulli measure from section 8.3 we see that the shifts

$$\begin{cases} \sigma : \prod_{n \in \mathbb{Z}} \{1, 2\} \to \prod_{n \in \mathbb{Z}} \{1, 2\} \text{ with probability vector } (\tfrac{1}{2}, \tfrac{1}{2}), \\ \sigma : \prod_{n \in \mathbb{Z}} \{1, 2\} \to \prod_{n \in \mathbb{Z}} \{1, 2, 3\} \text{ with probability vector } (\tfrac{1}{3}, \tfrac{1}{3}, \tfrac{1}{3}) \end{cases}$$

have entropies $\log 2$ and $\log 3$, respectively. Therefore they are *not* isomorphic.

REMARK. Let us consider a slightly different situation where we drop the assumption that ϕ is a bijection. That is, if we consider $(X_i, \mathcal{B}_i, \mu_i)$, for $i = 1, 2$, to be probability spaces then a *factor map* between $T_1 : X_1 \to X_1$ and $T_2 : X_2 \to X_2$ satisfies

(1) $\phi(X_1)$ is equal to X_2 (after removing a set of zero μ_2-measure, if necessary),
(2) ϕ is measurable,
(3) $\mu_1(\phi^{-1}B) = \mu_2(B)$ for $B \in \mathcal{B}_2$,
(4) $\phi \circ T_1 = T_2 \circ \phi$.

In this case it is easy to see that $h_\mu(T_1) \geq h_\mu(T_2)$.
We usually say that T_1 is an *extension* of T_2 or that T_2 is an *factor* of T_1.

8.9 Comments and references

Our development of entropy has followed the lines of Parry's treatment in [2]. It is possible to reduce some of the analysis if we accept working only with countable sigma-algebras [5].

References for some more advanced topics we have omitted include [3, §7.5] (Krieger's generator theorem), [3, §7.6] (Ornstein's isomorphism theorem), and [1, §10.7] (Keane-Smorodinsky finitary Isomorphism Theorem).

References

1. I. Cornfold, S. Fomin and Y. Sinai, *Ergodic Theory*, Springer, Berlin, 1982.
2. W. Parry, *Topics in Ergodic Theory*, C.U.P., Cambridge, 1981.
3. D. Rudolph, *Fundamentals of Measurable Dynamics*, O.U.P., Oxford, 1990.
4. Y. Sinai, *Topics in Ergodic Theory*, P.U.P., Princeton N.J., 1994.
5. P. Walters, *An Introduction to Ergodic Theory*, Springer, Berlin, 1982.

CHAPTER 9

ERGODIC MEASURES

In this chapter we shall consider the stronger property of ergodicity for an invariant probability measure μ. This property is more appropriate (amongst other things) for understanding the "long term" average behaviour of a transformation.

9.1 Definitions and characterization of ergodic measures

DEFINITION. Given a probability space $(X, \mathcal{B}, \mu)$, a transformation $T : X \to X$ is called *ergodic* if for every set $B \in \mathcal{B}$ with $T^{-1}B = B$ we have that either $\mu(B) = 0$ or $\mu(B) = 1$.

Alternatively we say that μ is T-ergodic.

The following lemma gives a simple characterization in terms of functions.

LEMMA 9.1. *T is ergodic with respect to μ iff whenever $f \in L^1(X, \mathcal{B}, \mu)$ satisfies $f = f \circ T$ then f is a constant function.*

PROOF. This is an easy observation using indicator functions. ∎

9.2 Poincaré recurrence and Kac's theorem

We begin with one of the most fundamental results in ergodic theory.

THEOREM 9.2 (POINCARÉ RECURRENCE THEOREM). *Let $T : X \to X$ be a measurable transformation on a probability space $(X, \mathcal{B}, \mu)$. Let $A \in \mathcal{B}$ have $\mu(A) > 0$; then for almost points $x \in A$ the orbit $\{T^n x\}_{n \geq 0}$ returns to A infinitely often.*

PROOF. Let $F = \{x \in A : T^n x \notin A, \forall n \geq 1\}$, then it suffices to show that $\mu(F) = 0$.

Towards this end, we first observe that $T^{-m}F \cap T^{-n}F = \emptyset$ when $n > m$, say. If this were not the case and $w \in T^{-m}F \cap T^{-n}F$ then $T^m w \in F$ and $T^{n-m}(T^m w) \in F \subset A$, which contradicts the definition of F.

Thus since the sets $\{T^{-n}F\}_{n\geq 0}$ are disjoint we see that

$$\sum_{n=0}^{\infty} \mu(T^{-n}F) = \mu(\cup_{n=0}^{\infty}T^{-n}F) \leq \mu(X) = 1$$

and then because μ is T-invariant $\mu(F) = \mu(T^{-1}F) = \ldots = \mu(T^{-n}F) = \ldots$
so we can only have that $\mu(F) = 0$.   ∎

DEFINITION. Let $n_A : A \rightarrow \mathbb{Z}^+ \cup \{+\infty\}$ be the first return time i.e.
$n_A(x) > 0$ is the smallest value for which $T^{n_A(x)}x \in A$.

By Theorem 9.2 $n_A(x)$ is finite almost everywhere. The next theorem
shows that when μ is an ergodic measure then the *average* return time to A
can be calculated explicitly.

THEOREM 9.3 (KAC'S THEOREM). *Let $T : X \rightarrow X$ be an ergodic trans-
formation on a probability space $(X, \mathcal{B}, \mu)$. Let $A \in \mathcal{B}$ have $\mu(A) > 0$ then we
define the return time function $n_A : A \rightarrow \mathbb{Z}^+ \cup \{\infty\}$ (which is finite, almost
everywhere). The average return time (with respect to the induced probability
measure μ_A) is*

$$\int_A n_A(x)d\mu_A(x) = \frac{1}{\mu(A)}.$$

PROOF. By definition of μ_A it is equivalent to show that $\int_A n_A(x)d\mu(x) = 1$. It is useful to define the following sets.

(a) For each $n \geq 1$ we define $A_n = \{x \in A : n(x) = n\}$, and write $A = \cup_{n\geq 1}A_n$ (with $A_i \cap A_j = \emptyset$ for $i \neq j$). In particular, $\sum_{n=1}^{\infty} \mu(A_n) = \mu(A)$.

(b) For $n \geq 1$ we define $B_n = \{x \in X : T^j x \notin A$ for $1 \leq j \leq n-1, T^n x \in A\}$. The sets B_n are disjoint (i.e. $B_i \cap B_j = \emptyset$ for $i \neq j$) and by ergodicity $X = \cup_{n\geq 1}B_n$ (since $\cup_{n\geq 1}B_n \supset \cup_{n\geq 1}T^{-n}A \supset A$) so that $\sum_{n=1}^{\infty} \mu(B_n) = 1$.

We can rewrite

$$\int_A n_A(x)d\mu(x) = \sum_{k=1}^{\infty} k\mu(A_k) = \sum_{k=1}^{\infty} \left(\sum_{n=k}^{\infty} \mu(A_n)\right);$$

then if we can show that $\sum_{n=k}^{\infty} \mu(A_n) = \mu(B_k)$ this will complete the proof.

When $k = 1$ we have from the definitions that $B_1 = T^{-1}A$ and so
$\sum_{n=1}^{\infty} \mu(A_n) = \mu(A) = \mu(B_1)$, as required. For $k > 1$ we can proceed
by induction. We can partition $T^{-1}B_k = B_{k+1} \cup T^{-1}A_k$ (where $T^{-1}A_k = T^{-1}B_k \cap T^{-1}A$ and $B_{k+1} = T^{-1}B_k \cap (X - T^{-1}A)$). Thus $\mu(T^{-1}B_k) = \mu(B_k) = \mu(B_{k+1}) + \mu(T^{-1}A_k) = \mu(B_{k+1}) + \mu(A_k)$ and using the inductive
hypothesis we see that $\mu(B_{k+1}) = \mu(B_k) - \mu(A_k) = \sum_{n=k+1}^{\infty} \mu(A_n)$. This
completes the inductive step and the proof.   ∎

9.3 Existence of ergodic measures

When $T : X \to X$ is a continuous map on a compact metric space there is a very simple relationship between ergodic measures and invariant measures which we can now describe.

Let $\mathcal{M}$ denote the set of invariant probability measures on X. There is a natural topology on this space called the *weak-star* topology, i.e. the weakest topology such that a sequence $\mu_n \in \mathcal{M}$ converges to $\mu \in \mathcal{M}$ iff $\forall f \in C^0(X)$, $\int f d\mu_n \to \int f d\mu$.

The following properties of $\mathcal{M}$ are well-known (and easily checked):

(i) $\mathcal{M}$ is convex (i.e. if $\mu_1, \mu_2 \in \mathcal{M}$ and $0 < \alpha < 1$, then $\alpha\mu_1 + (1-\alpha)\mu_2 \in \mathcal{M}$);

(ii) the set $\mathcal{M}$ is compact (in the weak-star topology) [5, Theorem 6.10].

LEMMA 9.4. *The extremal points in the convex set $\mathcal{M}$ are ergodic measures (i.e. $\mu \in \mathcal{M}$ is ergodic if whenever $\exists \mu_1, \mu_2 \in \mathcal{M}$ and $0 < \alpha < 1$ with $\mu = \alpha\mu_1 + (1 - \alpha)\mu_2$ then $\mu_1 = \mu_2$).*

The converse is also true, but we shall not require it.

PROOF. If μ is not ergodic then we can find $B \in \mathcal{B}$ with $T^{-1}B = B$ and $0 < \mu(B) < 1$. But for any set $A \in \mathcal{B}$ we can write $A = (A\cap B)\cup(A\cap(X-B))$ and thus

$$\mu(A) = \mu\left((A \cap B) \cup (A \cap (X - B))\right)$$

$$= \mu(B)\left(\frac{\mu(A \cap B)}{\mu(B)}\right) + \mu(X - B)\left(\frac{\mu(A \cap (X - B))}{\mu(X - B)}\right)$$

$$= \alpha\mu_1(A) + (1 - \alpha)\mu_2(A)$$

where $\alpha = \mu(B)$ and $\mu_1(A) = \left(\frac{\mu(A\cap B)}{\mu(B)}\right)$, $\mu_2(A) = \left(\frac{\mu(A\cap(X-B))}{\mu(X-B)}\right)$. This shows that $\mu = \alpha\mu_1 + (1 - \alpha)\mu_2$. ∎

PROPOSITION 9.5 (EXISTENCE OF ERGODIC MEASURES). *Let X be a compact metric space and $\mathcal{B}$ be the Borel sigma-algebra. Given any continuous map $T : X \to X$ there exists at least one T-ergodic probability measure μ.*

PROOF. Choose a dense set of functions $f_k \in C^0(X)$, $k \geq 0$. Since the map $\mu \to \int f_0 d\mu$ is continuous on $\mathcal{M}$ there exists by (weak-star) compactness at least one $\nu \in \mathcal{M}$ such that $\int f_0 d\nu = \sup_{\mu \in \mathcal{M}}\{\int f_0 d\mu\}$. We let

$$\mathcal{M}_0 = \left\{\nu \in \mathcal{M} : \int f_0 d\nu = \sup_{\mu \in \mathcal{M}}\left\{\int f_0 d\mu\right\}\right\};$$

then clearly $\mathcal{M}_0$ is non-empty and closed. Similarly, define

$$\mathcal{M}_1 = \left\{\nu \in \mathcal{M}_0 : \int f_1 d\nu = \sup_{\mu \in \mathcal{M}_0}\left\{\int f_1 d\mu\right\}\right\}$$

and the same reasoning shows that $\mathcal{M}_1 \subset \mathcal{M}_0 \subset \mathcal{M}$ is non-empty and closed.

Proceeding inductively we define

$$\mathcal{M}_k = \left\{ \nu \in \mathcal{M}_{k-1} : \int f_k d\nu = \sup_{\mu \in \mathcal{M}_{k-1}} \left\{ \int f_k d\mu \right\} \right\}$$

and arrive at a nested sequence $\mathcal{M} \supset \mathcal{M}_0 \supset \mathcal{M}_1 \supset M_2 \supset \ldots \supset \mathcal{M}_k \supset \ldots$. Since the sets are all closed in $\mathcal{M}$ (and hence compact) we have that the intersection is non-empty. Assume $\mu \in \cap_{k \in \mathbb{Z}^+} \mathcal{M}_k$. We want to show that μ is ergodic by showing that it is an extreme point in $\mathcal{M}$.

Assume that μ can be written as an affine combination $\mu = \alpha \mu_1 + (1-\alpha)\mu_2$ (with $0 < \alpha < 1$); then to show that μ is ergodic we need to show that $\mu_1 = \mu_2$. Thus it suffices to show that for every $f_k \in C^0(X)$ we have that $\int f_k f \mu_1 = \int f_k f \mu_2$ (since the set f_k is dense).

We begin with $k = 0$ and observe that by assumption $\int f_0 d\mu = \alpha \int f_0 d\mu_1 + (1 - \alpha) \int f_0 d\mu_2$. Since $\mu \in \mathcal{M}_0$ we see that $\sup_{m \in \mathcal{M}} \{ \int f_0 dm \} = \int f_0 d\mu$ implies that $\int f_0 d\mu_1 = \int f_0 d\mu_2 = \sup_{m \in \mathcal{M}} \{ \int f_0 dm \}$. We thus conclude

(1) the first identity $\int f_0 d\mu_1 = \int f_0 d\mu_2$ is proved.

(2) $\mu_1, \mu_2 \in \mathcal{M}_0$.

Continuing inductively, we establish that for arbitrary $k \geq 0$ we have $\int f_k d\mu_1 = \int f_k d\mu_2$ and $\mu_1, \mu_2 \in \mathcal{M}_k$. This completes the proof (i.e. $\mu_1 = \mu_2$ and μ is an extremal measure).   ∎

REMARK. The following facts are easy to check.

(3) If ν, μ are distinct T-ergodic measures then $\nu \perp \mu$.

(4) If μ is ergodic then it is an extremal measure in $\mathcal{M}$. (The converse to Lemma 9.4.)

Since $\mathcal{M}$ is a compact convex metric space there is a general theorem of Choquet that says *every invariant measure $\mu \in \mathcal{M}$ can be written as a convex combination of extremal measures in $\mathcal{M}$*. More precisely, we can find a measure $\rho = \rho_\mu$ on the space $\mathcal{M}$ (with respect to the Borel sigma-algebra associated to the weak-star topology) such that

(1) for any function $f \in C^0(X)$ we have

$$\int f d\mu = \int_{\mathcal{M}} \left(\int f d\nu \right) d\rho(\nu).$$

(2) $\rho(\{ \nu : \nu \text{ is extremal} \}) = 1$.

9.4 Some basic constructions in ergodic theory

In this final section of chapter 9 we shall describe two basic constructions in ergodic theory.

9.4.1. Skew products. Let $T : X \to X$ be a measure preserving transformation of a probability space $(X, \mathcal{B}, \mu)$. Let $(G, \mathcal{B})$ be a compact Lie group with the Borel sigma-algebra $\mathcal{B}$. We can consider the product space $X \times \mathbb{R}/\mathbb{Z}$ with the product sigma-algebra $\mathcal{A}$.

DEFINITION. Given a measure preserving transformation of $T : X \to X$ and a measurable map $\phi : X \to G$ we define a *skew product* to be the transformation $S : X \times G \to X \times G$ defined by $S(x, g) = (Tx, \phi(x)g)$. Given any T-invariant probability measure μ we can associate the S-invariant measure ν defined by $d\nu = d\mu \times dt$.

A simple example is the following.

EXAMPLE. Let $T : \mathbb{R}/\mathbb{Z} \to \mathbb{R}/\mathbb{Z}$ be given by $T(x) = x + \alpha \pmod 1$ for some $\alpha \in \mathbb{R}$. Let $G = \mathbb{R}/\mathbb{Z}$ and we define $\phi : \mathbb{R}/\mathbb{Z} \to \mathbb{R}/\mathbb{Z}$ by $\phi(x) = x \pmod 1$ (i.e. the identity map). The associated skew product is then the map $S : \mathbb{R}^2/\mathbb{Z}^2 \to \mathbb{R}^2/\mathbb{Z}^2$ given by $S(x, y) = (x + \alpha, x + y) \pmod 1$.

9.4.2. Induced transformations and Rohlin towers. Assume that $T : X \to X$ is a measurable transformation on a measurable space $(X, \mathcal{B})$. Assume that $A \subset X$ with $A \in \mathcal{B}$.

DEFINITION. The transformation $T_A : A \to A$ defined by $T_A(x) = T^{n(x)}x$ is called the *induced transformation* on A. We denote by $\mathcal{B}_A = \{B \cap A : B \in \mathcal{B}\}$ the restriction of the sigma-algebra $\mathcal{B}$ to A.

If μ is a T-invariant sigma-finite measure on $(X, \mathcal{B})$ and $\mu(A) > 0$ then we can define a T_A-invariant measure μ_A on $(A, \mathcal{B}_A)$ by $\mu_A(B) = \frac{\mu(A \cap B)}{\mu(A)}$.

EXAMPLE (CONTINUED FRACTION TRANSFORMATION). Consider the case where $(X, \mathcal{B})$ is the positive half-line $\mathbb{R}^+ = (0, +\infty)$ with the Borel sigma-algebra. We define a transformation $T : \mathbb{R}^+ \to \mathbb{R}^+$ by

(1) $Tx = x - 1$ if $x \in [1, +\infty)$, and
(2) $Tx = \frac{1}{x}$ if $x \in (0, 1)$.

We can consider the induced transformation $T_A : A \to A$ on the interval $A = (0, 1]$ defined by $T_A x = \frac{1}{x} - \left[\frac{1}{x}\right]$.

The measure μ_A defined by $\mu_A(B) = \frac{1}{\log 2} \int_B \frac{1}{1+x} dx$ is T_A-invariant.

REMARK. We need not be too careful about the definition of T and T_A on a countable set of points since they have zero measure.

Consider an (ergodic) transformation $T : X \to X$ on a probability space $(X, \mathcal{B}, \mu)$ and let $A \in \mathcal{B}$ have $\mu(A) > 0$.

DEFINITION. We can define a space

$$A^{n_A} = \{(x, k) \in A \times \mathbb{Z}^+ : 0 \leq k \leq n_A(x)\},$$

where we identify $(x, n_A(x)) \sim (T^{n_A(x)}x, 0)$, and introduce the product sigma-algebra $\bar{\mathcal{B}}$ (i.e. the smallest sigma-algebra containing the products of sets in $\mathcal{B}_A$ and $\mathcal{B}_{\mathbb{Z}^+}$).

We define a probability measure on the space A^{n_A} by $\nu = \frac{\mu \times dn}{\int n_A d\mu}$ (where dn corresponds to the usual counting measure on $\mathbb{Z}^+$).

Finally, we define a transformation $T_A^{n_A} : A^{n_A} \to A^{n_A}$ by

(1) $T_A^{n_A}(x, k) = (x, k+1)$ if $0 \le k < n_A(x)$, and
(2) $T_A^{n_A}(x, n_A(x)) = T_A^{n_A}(T_A x, 0)) = (T_A x, 1)$.

This construction is called the *Rohlin tower* over A.

(N.B. A Rohlin tower is the converse process to induced transformations. We reproduce the original transformation on X from the induced transformation on A.) The following lemma tells us the Rohlin tower is a good model of the original transformation.

LEMMA 9.6. *The map* $\phi : (A^{n_A}, \bar{\mathcal{B}}, \nu) \to (X, \mathcal{B}, \mu)$ *defined by* $\phi(x, k) = T^k(x)$ *is measurable and satisfies the following:*

(1) ϕ *is a bijection (almost everywhere);*
(2) $\forall B \in \mathcal{B}$ *we have that* $\nu(\phi^{-1}B) = \mu(B)$; *and*
(3) $\phi T_A^{n_A} = T\phi$ *(almost everywhere).*

PROOF. The result follows almost immediately from the definitions. ∎

REMARK. The map ϕ is an *isomorphism* which implies that from the point of view of ergodic theory the transformations T and $T_A^{n_A}$ are the same.

9.4.3 Natural extensions. Given a non-invertible map $T : (X, \mathcal{B}, \mu) \to (X, \mathcal{B}, \mu)$ there is a natural way of associating to it an invertible transformation $\hat{T} : (\hat{X}, \hat{\mathcal{B}}, \hat{\mu}) \to (\hat{X}, \hat{\mathcal{B}}, \hat{\mu})$ with similar dynamical properties.

We define

$$\hat{X} = \{(x_n)_{n \in \mathbb{Z}^+} \in \prod_{n \in \mathbb{Z}^+} X : T(x_n) = x_{n+1}, n \ge 0\}$$

and associate the sigma-algebra generated by the sets

$$B_m := \left\{(x_n)_{n \in \mathbb{Z}^+} \in \hat{X} : x_m \in B\right\} \text{ for } B \in \mathcal{B} \text{ and } m \in \mathbb{Z}^+.$$

We next define a probability measure $\hat{\mu}$ on $\hat{\mathcal{B}}$ by $\hat{\mu}(B_m) = \mu(B)$. Finally, we define the (invertible) transformation $\hat{T} : X \to X$ by

$$\hat{T}(x_0, x_1, x_2, \dots) = (Tx_0, x_0, x_1, x_2, \dots).$$

It is easy to see from the construction that $\hat{T}$ is measurable and preserves the probability measure $\hat{\mu}$.

DEFINITION. We call $\hat{T} : \hat{X} \to \hat{X}$ the *natural extension* of X.

There is a canonical map $\pi : \hat{X} \to X$ defined by $\pi\left((x_n)_{n \in \mathbb{Z}^+}\right) = X$. The natural extension $\hat{T}$ has the following properties:

(i) $\hat{T}$ is an extension of T in the sense that $\pi \circ \hat{T} = T \circ \pi$; and

(ii) if we denote by $\hat{\mathcal{B}}^+ \subset \hat{\mathcal{B}}$ the sub-sigma-algebra generated by sets $\{\pi^{-1}(B) : B \in \mathcal{B}\}$ then

$$\ldots \subset \hat{T}^{-1}\hat{\mathcal{B}}^+ \subset \hat{\mathcal{B}}^+ \subset T\hat{\mathcal{B}}^+ \subset \ldots \subset \cup_{n \in \mathbb{Z}^+} T^n \hat{\mathcal{B}}^+ = \mathcal{B}.$$

REMARK. In fact, any transformation satisfying (i) and (ii) will be isomorphic to the natural extension as we have defined it above [3].

EXAMPLE (SUBSHIFTS OF FINITE TYPE). Let $\sigma : X_A^+ \to X_A^+$ be a (one-sided) subshift of finite type, defined by the $k \times k$ matrix A. Relative to a Markov measure, say, its natural extension is the shift $\sigma : X \to X$.

9.5 Comments and references

More can be found on ergodic measures in [1], [2] and [5].

Important applications of ergodic theory beyond the scope of these notes are Mostow's rigidity theorem [4] and the Margulis super-rigidity theorem [6, §5.1].

The skew product example in subsection 9.4.1 was used by Furstenburg to give a simple proof of a result on diophantine approximation due to Hardy and Littlewood [2].

References

1. P. Billingsley, *Ergodic Theory and Information*, Wiley, New York, 1965.
2. W. Parry, *Topics in Ergodic Theory*, C.U.P., Cambridge, 1981.
3. V. Rohlin, *Exact endomorphisms of lebesgue space*, Amer. Math. Soc. Transl. (2) **39** (1964), 1-36.
4. W. Thurston, *Topology and geometry of three manifolds*, unpublished notes, Princeton University N.J., Princeton, 1978.
5. P. Walters, *An Introduction to Ergodic Theory*, Springer, New York, 1989.
6. R. Zimmer, *Ergodic Theory and Semi-simple Groups*, Birkhäuser, Basel, 1984.

CHAPTER 10

ERGODIC THEOREMS

In this chapter we shall describe the ergodic theorems and some of their applications. These simple but elegant theorems are important in many other areas.

Let f be a function which is an observable for a physical quantity. One of the main themes in ergodic theory is to study the asymptotic behaviour of their time evolution $\{f \circ T^k\}_{k \in Z^+}$. Under the ergodic hypothesis, their averages $\frac{1}{N} \sum_{k=0}^{N-1} f \circ T^k$ converge to the space average $\int f d\mu$. This property also implies the well-known law of large numbers, which is a key concept in statistics (i.e. the distribution of the long term average converges to the Dirac measure supported on $\int f d\mu$).

10.1 The von Neumann ergodic theorem. Let $(X, \mathcal{B}, \mu)$ be a measure space with μ a probability measure, and assume that $T : X \to X$ preserves μ. In general, ergodic theorems describe how the averages of functions $f : X \to \mathbb{R}$ along the orbits $\{T^n x\}_{n=0}^{\infty}$ behave for (almost all) points $x \in X$. The simplest theorem of this type is the following "mean" ergodic theorem.

THEOREM 10.1 (VON NEUMANN MEAN ERGODIC THEOREM). *Let $(X, \mathcal{B}, \mu)$ be a probability space, and assume that $T : X \to X$ preserves μ. Then for $f, g \in L^2(X, \mathcal{B}, \mu)$ the averages*

$$\frac{1}{N} \sum_{n=0}^{N-1} \int f(T^n x) g(x) d\mu(x)$$

converge as $N \to +\infty$. If T is ergodic then

$$\frac{1}{N} \sum_{n=0}^{N-1} \int f(T^n x) g(x) d\mu(x) \to \int f d\mu \int g d\mu$$

as $N \to +\infty$.

(Cf. [6, §2], [8, pp. 23-27].)

PROOF. Consider the subspace $I = \{f \in L^2(X, \mathcal{B}, \mu) : fT = f\}$ of T - invariant functions. (If T is ergodic then I consists of constant functions.) Consider too the subspace

$$B = \{f \in L^2(X, \mathcal{B}, \mu) : \exists h \in L^2(X, \mathcal{B}, \mu) \text{ with } f = hT - h\}$$

of *coboundaries.*

On each of these spaces the averages are easier to study.

On the first space I we see that for $f_1 \in I$ and any $N \geq 1$

$$\frac{1}{N} \sum_{n=0}^{N-1} f_1(T^n x) = f_1(x) \in L^2(X, \mathcal{B}, \mu).$$

On the second space B we see that if $f_2 = h \circ T - h \in B$ then

$$\frac{1}{N} \sum_{n=0}^{N-1} f_2(T^n x) = \frac{1}{N} \left(h(T^N x) - h(x) \right).$$

By the Cauchy-Schwarz inequality we have that

$$\frac{1}{N} | \int \left(h(T^N x) - h(x) \right) g d\mu(x)| \leq \frac{1}{N} \left(\left| \int h \circ T^N \cdot g d\mu \right| + \left| \int h \cdot g d\mu \right| \right)$$

$$\leq \frac{1}{N} \left(||h \circ T^n||_2 ||g||_2 + ||h||_2 ||g||_2 \right)$$

$$= \frac{2}{N} ||h||_2 ||g||_2$$

which converges to zero as $N \to +\infty$.

If we can write $f = f_1 + f_2$ with $f_1 \in I$ and $f_2 \in B$ then

$$\frac{1}{N} \sum_{n=0}^{N-1} \int f(T^n x) g(x) d\mu(x)$$

$$= \frac{1}{N} \sum_{n=0}^{N-1} \int f_1(T^n x) g(x) d\mu(x) + \frac{1}{N} \sum_{n=0}^{N-1} \int f_2(T^n x) g(x) d\mu(x)$$

$$= \int f_1(x) g(x) d\mu(x) + \frac{1}{N} \sum_{n=0}^{N-1} \int f_2(T^n x) g(x) d\mu(x)$$

and therefore

$$\frac{1}{N} \sum_{n=0}^{N-1} \int f(T^n x) g(x) d\mu(x) \to \int f_1(x) g(x) d\mu(x) = \int f(x) g(x) d\mu(x)$$

as $N \to +\infty$.

More generally, assume that for any $\epsilon > 0$ we can find $f_1 \in I$ and $f_2 \in B$; so then $\int |f - (f_1 + f_2)|^2 d\mu < \epsilon$ then by the Cauchy-Schwarz inequality

$$\frac{1}{N} \sum_{n=0}^{N-1} \int (f - (f_1 + f_2))(T^n x) g(x) d\mu(x)$$

$$\leq \frac{1}{N} \sum_{n=0}^{N-1} \|(f - (f_1 + f_2)) T^n\|_2 \|g\|_2$$

$$= \frac{1}{N} \sum_{n=0}^{N-1} \|(f - (f_1 + f_2))\|_2 \|g\|_2$$

$$\leq \epsilon \|g\|_2$$

(where to get from the second to third line we use the property that T preserves μ to write $\|h \circ T^n\|_2^2 = \int |h \circ T^n| d\mu = \int |h| \circ T^n d\mu = \int |h| d\mu = \|h\|_2^2$). Therefore the limit (points) of the sequence

$$\frac{1}{N} \sum_{n=0}^{N-1} \int f(T^n x) g(x) \mu(x)$$

lie in the interval $\left(\int f d\mu \int g d\mu - \epsilon, \int f d\mu \int g d\mu + \epsilon \right)$. Since $\epsilon > 0$ can be chosen arbitrarily small we see that

$$\frac{1}{N} \sum_{n=0}^{N-1} \int f(T^n x) g(x) d\mu(x) \to \int f d\mu \int g d\mu$$

as $N \to +\infty$.

So to prove the theorem it suffices to show that $L^2(X, \mathcal{B}, \mu) = I + \mathrm{cl}(B)$. To see this, assume that $f \perp cl(B)$, i.e. $\int f \cdot f_2 d\mu = 0, \forall f_2 \in \mathrm{cl}(B)$ or in particular $\int f \cdot h \circ T d\mu = \int f \cdot h d\mu, \forall h \in L^2(X, \mathcal{B}, \mu)$. Taking $h = f$ gives us that

$$\int |f|^2 d\mu = \int f \circ T \cdot f d\mu. \tag{10.1}$$

We need to show that $f \in I$ (i.e. $f \circ T = f$) and to this end we can write

$$\int |f \circ T - f|^2 d\mu = \int |f \circ T|^2 d\mu + \int |f|^2 d\mu - 2 \int f \circ T.f d\mu$$

$$= 2\left(\int |f|^2 d\mu - \int f \circ T.f d\mu \right)$$

$$= 0$$

where we use (10.1) and that $\int |f \circ T|^2 d\mu = \int |f|^2 d\mu$. This shows that $f \circ T = f$ a.e. (i.e. $f \in I$).

This completes the proof.

10.2 The Birkhoff theorem (for ergodic measures)

Let (X, B, μ) be a probability space, and assume that the transformation $T : X \to X$ preserves μ. The Birkhoff "individual" ergodic theorem gives a strong type of ergodic theorem in that it describes the average of functions along individual "typical" orbits.

We first prove the theorem under the additional assumption that μ is ergodic, and then for arbitrary invariant measures.

THEOREM 10.2 (BIRKHOFF'S THEOREM (ERGODIC VERSION)). *Consider* $f \in L^1(X, B, \mu)$. *If the measure μ is ergodic then for almost all $x \in X$ we have that the averages*

$$\frac{1}{N} \sum_{n=0}^{N-1} f(T^n x) \to \int f d\mu$$

as $N \to +\infty$ (i.e. $\mu\{x \in X: \lim_{N \to \infty} \frac{1}{N} \sum_{n=0}^{N-1} f(T^n x) \neq \int f d\mu\} = 0$).

(Cf. [6 §1.2], [8, p. 30].)

PROOF. Assume for convenience that $\int f d\mu = 0$. (If this is not the case then replace f by $f - \int f d\mu$ in the two sides of the equality in the statement of the theorem.)

For any $\epsilon > 0$ we can define

$$E_\epsilon(f) = \{x \in X : \limsup_{N \to +\infty} \frac{1}{N} | \sum_{n=0}^{N-1} f(T^n x)| \geq \epsilon\}.$$

To prove the theorem it suffices to show that for each $\epsilon > 0$ we have that $\mu(E_\epsilon) = 0$. We do this via two lemmas.

We begin with the following estimate.

SUBLEMMA 10.2.1. $\mu(E_{2\epsilon}(f)) \leq \frac{\int |f| d\mu}{\epsilon}$.

PROOF. We first write $f = f_+ - f_-$ where $f_+, f_- \geq 0$.
We next define the sets

$$E_\epsilon^M(f_+) = \{x \in X : \exists 1 \leq N \leq M, \sum_{n=0}^{N-1} f_+(T^n x) \geq \epsilon N\}$$

and

$$E_\epsilon^M(f_-) = \{x \in X : \exists 1 \leq N \leq M, \sum_{n=0}^{N-1} f_-(T^n x) \geq \epsilon N\}$$

for any $M \geq 1$.

Next observe that for any $x \in X$ and any $P > M$ we have the lower bounds

$$\sum_{n=0}^{P-1} f_+(T^n x) \geq \epsilon \sum_{j=0}^{P-M} \chi_{E_\epsilon^M(f_+)}(T^j x)$$

and

$$\sum_{n=0}^{P-1} f_-(T^n x) \geq \epsilon \sum_{j=0}^{P-M} \chi_{E_\epsilon^M(f_-)}(T^j x)$$

where we bound $f_+(T^n x)$ or $f_-(T^n x)$ from below by ϵ or 0 as appropriate. Integrating each side of these identities gives that

$$\int \left(\sum_{n=0}^{P-1} f_+(T^n x) \right) d\mu(x) = P \int f_+ d\mu \geq (P-M)\epsilon\mu\left(E_\epsilon(f_+)\right)$$

and

$$\int \left(\sum_{n=0}^{P-1} f_-(T^n x) \right) d\mu(x) = P \int f_- d\mu \geq (P-M)\epsilon\mu\left(E_\epsilon(f_-)\right)$$

for all $M \geq 1$.

Letting $P \to +\infty$ gives that

$$\int f_+ d\mu \geq \epsilon\mu\left(E_\epsilon(f_+)\right) \quad \text{and} \quad \int f_- d\mu \geq \epsilon\mu\left(E_\epsilon(f_-)\right). \tag{10.2}$$

We finally see that

$$\mu\left(E_{2\epsilon}(f)\right) \leq \limsup_{M \to +\infty} \mu\left(E_\epsilon(f_+)\right) + \limsup_{M \to +\infty} \mu\left(E_\epsilon(f_-)\right)$$

$$\leq \frac{\int f_+ d\mu}{\epsilon} + \frac{\int f_- d\mu}{\epsilon}$$

$$= \frac{\int |f| d\mu}{\epsilon}.$$

This completes the proof of Sublemma 10.2.1.

∎

The inequality in Sublemma 10.2.1 is not effective unless we can control the size of $\int |f| d\mu$. The next sublemma allows us to do this (and this is also where the ergodicity of μ is used).

SUBLEMMA 10.2.2. *Let $\int f d\mu = 0$. Given $\delta > 0$ we can choose a function $h \in L^\infty(X, B, \mu)$ with $\int |f - (hT - h)| d\mu < \delta$.*

PROOF. We need to show that the subspace

$$E := \{hT - h : h \in L^\infty(X, \mathcal{B}, \mu)\}$$

is dense in the subspace

$$B_0 = \{f \in L^1(X, \mathcal{B}, \mu) : \int f d\mu = 0\}.$$

By the Hahn-Banach theorem it suffices to show that all linear functionals which vanish on E also vanish on B_0.

As is well-known, for each linear functional α on $L^1(X, \mathcal{B}, \mu)$ there exists $k \in L^\infty(X, \mathcal{B}, \mu)$ such that $\alpha(f) = \int f k d\mu$. Given a linear functional, and thus $k \in L^\infty(X, \mathcal{B}, \mu)$, the hypothesis that the linear functional vanishes on E implies that $\int (hT - h)k d\mu = 0$, $\forall h \in L^\infty(X, \mathcal{B}, \mu)$.

With the specific choice $h = k$ we get that $\int kT \cdot k d\mu = \int k^2 d\mu$ which implies that

$$\int (k \circ T - k)^2 d\mu = \int (k \circ T)^2 d\mu + \int (k)^2 d\mu - 2 \int k \circ T.k d\mu$$

$$= 2 \left(\int (k)^2 d\mu - \int k \circ T.k d\mu \right)$$

$$= 0.$$

In particular, $k \circ T = k$. By the ergodicity assumption this implies that k is constant. So if $f \in B_0$ then $\int f d\mu = 0$, which in turn implies that $\int f k d\mu = 0$ and so f is in the kernel of the functional, i.e. the functional vanishes on B_0.   ∎

We can assume without loss of generality that $f \in B_0$ (otherwise we replace f by $f - \int f d\mu$). For an arbitrary $\delta > 0$ we can use sublemma 10.2.2 to choose $h \in L^\infty(X, \mathcal{B}, \mu)$ with $\int |f - (hT - h)| d\mu < \delta$.

From the definitions we see that for arbitrary $\epsilon > 0$ we have $E_\epsilon(f) \subset E_{\frac{\epsilon}{2}}(f - (hT - h)) \cup E_{\frac{\epsilon}{2}}(hT - h)$ and therefore

$$\mu\left(E_\epsilon(f)\right) \le \mu\left(E_{\frac{\epsilon}{2}}(f - (hT - h))\right) + \mu\left(E_{\frac{\epsilon}{2}}(hT - h)\right). \tag{10.3}$$

However,

(1) since $\forall x \in X$

$$|\frac{1}{N} \sum_{n=0}^{N-1} (hT - h)(T^n x)| = \frac{1}{N}|h(T^N x) - h(x) \le \frac{2||h||_\infty}{N}$$

we see that $E_{\frac{\epsilon}{2}}((hT - h)) = \emptyset$ and so $\mu\left(E_{\frac{\epsilon}{2}}((hT - h))\right) = 0$.
(2) by sublemma 10.2.1 we have that

$$\mu\left(E_{\frac{\epsilon}{2}}(f - (hT - h))\right) \leq \frac{\int |f - (hT - h)| d\mu}{\epsilon/4} \leq \frac{4\delta}{\epsilon}.$$

Since $\delta > 0$ was arbitrary we deduce from part (2) that $\mu\left(E_{\frac{\epsilon}{2}}(f - (hT - h))\right)$
$= 0$ and so by (10.3) and part (1) that $\mu\left(E_\epsilon(f)\right) = 0$, $\forall \epsilon > 0$. This suffices
to prove the result.

∎

The following corollary shows that the Birkhoff Theorem gives a quantitative version of the Poincaré recurrence theorem.

COROLLARY 10.2.1. *Let (X, B, μ) be a probability space, and assume that
the transformation $T : X \to X$ preserves μ. The proportion of time spent by
almost all points in a subset $B \in \mathcal{B}$ is given by it measure $\mu(B)$, i.e.*

$$\lim_{N \to +\infty} \frac{1}{N} \mathrm{Card}\{0 \leq n \leq N - 1 : T^n x \in B\} = \mu(B)$$

for almost all points $x \in X$.

PROOF. This follows from applying the ergodic theorem to the indicator
function on the set A.

∎

EXAMPLE (LACK OF CONVERGENCE ON A SET OF ZERO MEASURE). Consider the map $T : \mathbb{R}/\mathbb{Z} \to \mathbb{R}/\mathbb{Z}$ defined by $T(x) = 2x \pmod 1$ and the usual
Haar-Lebesgue measure μ. Consider any continuous function $f : \mathbb{R}/\mathbb{Z} \to \mathbb{R}$
such that $f(0) \neq \int f d\mu$. Clearly,

$$\frac{1}{N} \sum_{n=0}^{N-1} f(T^n 0) = f(0)$$

since $T^n 0 = 0$, for all $n \geq 0$. But since $f(0) \neq \int f d\mu$ we see that at the point
0 the sequence does *not* converge to the integral.

REMARK. Given a flow $\phi_t : X \to X$ and an ergodic measure μ there is
a corresponding version of the ergodic theorem. For $f \in L^1(X, \mathcal{B}, \mu)$ the
averages $\frac{1}{T} \int_0^T f(\phi_t x) dt \to \int f(x) d\mu(x)$ as $T \to +\infty$ for almost all points
$x \in X$.

10.3 Applications of the ergodic theorems

In this section we shall consider some of the beautiful applications of the Birkhoff ergodic theorem.

Application 1 (Normal numbers). Let $k \geq 2$ be a positive integer. For any $0 < x < 1$ we can expand

$$x = \frac{i_1}{k} + \frac{i_2}{k^2} + \frac{i_3}{k^3} + \cdots$$

where $i_1, i_2, i_3, \ldots \in \{0, 1, \ldots, k-1\}$. (This expansion may not be unique.) When $k = 10$ this is the usual decimal expansion.

We say that x is normal to base k if for each $i \in \{0, 1, \ldots, k-1\}$ that term occurs in the expansion for x with density $\frac{1}{k}$, i.e. $\lim_{N \to +\infty} \frac{1}{N} \text{Card}\{1 \leq n \leq N : i_n = i\} = \frac{1}{k}$.

The transformation $T : [0, 1) \to [0, 1)$ defined by $Tx = kx \pmod 1$ preserves Lebesgue measure μ and is μ-ergodic. Moreover, we see that $T^n(x) \in \left(\frac{i_n}{k}, \frac{i_n+1}{k}\right)$ and so by Theorem 10.2 we have the following.

PROPOSITION 10.3. *For almost all $0 < x < 1$, x is normal to any base $k \geq 2$.*

Application 2 (First digits of powers of 2). Consider the sequence $2, 4, 8, 16, 32, 64, 128, \ldots, 2^k$ and consider the sequence of first digits $2, 4, 8, 1, 3, 6, 1, \ldots, r_k \in \{0, 1, \ldots, 9\}$.

PROPOSITION 10.4. *The frequency of the occurrence of the symbol k is given by $\log_{10}\left(1 + \frac{1}{k}\right)$, i.e.* $\lim_{N \to +\infty} \frac{1}{N} \text{Card}\{1 \leq n \leq N : r_n = k\} = \log_{10}\left(1 + \frac{1}{k}\right)$.

For example, the frequency of the occurrence of the symbol 7 is given by $\log_{10}\left(1 + \frac{1}{7}\right)$.

To prove Proposition 10.4, consider the distribution of the first digits of $\{2^n\}$. The first digit of 2^n equals k iff $k \cdot 10^r \leq 2^n < (k+1) \cdot 10^r$ for some $r \geq 0$. (Equivalently, $n \cdot \log_{10} 2 \in [\log_{10}(k+1), \log_{10} k) \pmod 1$).

But since $a = \log_{10} 2$ is irrational the map $Tx = x + a$ is ergodic with respect to Lebesgue measure. From the ergodic theorem it is easy to deduce that the proportion of the orbit $T^n 0$ spent in the interval $[\log_{10}(k+1), \log_{10} k)$ is equal to its length $|\log_{10}(k+1) - \log_{10} k| = \log_{10}(1 + \frac{1}{k})$.

Application 3 (Continued fractions). The continued fraction transformation $T : (0, 1) \to (0, 1)$ defined by $Tx = \frac{1}{x} - \left[\frac{1}{x}\right]$ preserves the *Gauss measure*

$$\mu(B) = \frac{1}{\log 2} \int_B \frac{dx}{1 + x}$$

where $B \in \mathcal{B}$ is in the Borel sigma-algebra.

In particular, for any $B \in \mathcal{B}$, if $l(B)$ denotes its measure with respect to the Haar-Lebesgue measure then

$$\frac{1}{2\log 2}l(B) \leq \mu(B) \leq \frac{1}{\log 2}l(B)$$

(since $\frac{1}{2} \leq \frac{1}{1+x} \leq 1$). Consider the (measurable) partition of $(0,1)$ by the intervals of the form $I_k := (\frac{1}{k+1}, \frac{1}{k})$ for $k \geq 1$. For any $n \geq 1$ we can define a finer partition into disjoint intervals by $I_{k_0} \cap T^{-1}I_{k_1} \cap \ldots \cap T^{-n}I_{k_n}$, where $k_0, k_1, \ldots, k_n \geq 1$.

We want to show that μ is ergodic. As is easily seen,

(i) $|T'(x)| > 1$ and $|(T^2)'(x)| \geq 4$, for any $x \in (0,1)$,
(ii) there exists $C > 0$ such that $|\frac{T''(x)}{T'(x)}| \leq C, \forall 0 < x < 1$.

The maps $T^{n+1} : I_{k_0} \cap T^{-1}I_{k_1} \cap \ldots \cap T^{-n}I_{k_n} \to [0,1]$ have inverses which we denote by $\psi_{k_0 \ldots k_n} : [0,1] \to I_{k_0} \cap T^{-1}I_{k_1} \cap \ldots \cap T^{-n}I_{k_n}$ and we see that

$$|\frac{\psi'_{k_0 \ldots k_n}(x)}{\psi'_{k_0 \ldots k_n}(y)}| \leq \prod_{i=0}^{n}\left(1 + \frac{C}{2^i}|x-y|\right) \leq D, \quad \forall x,y \in (0,1).$$

Fix a Borel-measurable set $E \in \mathcal{B}$ then by the change of variables formula we get the following two inequalities:

(a) $l\left(E \cap I_{k_0} \cap T^{-1}I_{k_1} \cap \ldots \cap T^{-n}I_{k_n}\right) = l\left(\psi_{k_0 \ldots k_n}(T^n E)\right)$
$= \int_{T^n E} |\psi'_{k_0 \ldots k_n}(x)|dx \geq l\left(T^n E\right)\left(\inf_{x \in [0,1]} |\psi'_{k_0 \ldots k_n}(x)|\right);$
(b) $l\left(I_{k_0} \cap T^{-1}I_{k_1} \cap \ldots \cap T^{-n}I_{k_n}\right) \leq \sup_{x \in [0,1]}\left(\psi'_{k_0 \ldots k_n}(x)\right).$

From the definitions we have

(c) $l(T^n E) \geq \log 2 \cdot \mu(T^n E) = \log 2\mu(T^{-n}T^n E) \geq \log 2\mu(E) \geq \frac{1}{2}l(E).$

In particular, we see that

$$l(E \cap I_{k_0} \cap T^{-1}I_{k_1} \cap \ldots \cap T^{-n}I_{k_n}) \geq \frac{1}{2D}l(E) \cdot l(I_{k_0} \cap T^{-1}I_{k_1} \cap \ldots \cap T^{-n}I_{k_n}).$$

Thus we see that for $E \in \mathcal{B}$ for which $T^{-1}E = E$ and any $\epsilon > 0$ we can choose n large and a cover for the complement $([0,1] - E) \subset \cup_i I_i$ by intervals I_i of the form $I_{k_0} \cap T^{-1}I_{k_1} \cap \ldots \cap T^{-n}I_{k_n}$ such that $\sum_i l(I_i) \leq$

$l([0,1] - E) + \epsilon = 1 - l(E) + \epsilon.$ Then

$$l(E) \cdot l([0,1] - E)$$

$$\leq l(E) \left(\sum_i l(I_i) + \epsilon \right)$$

$$\leq 2D \left(\sum_i l(E \cap I_i) \right) + \epsilon$$

$$= 2D \left(\sum_i l(E \cap I_i) + \left(\sum_i l(I_i \cap ([0,1] - E)) - l([0,1] - E) \right) \right) + \epsilon$$

$$= 2D \left(\sum_i l(I_i) - l([0,1] - E) \right) + \epsilon \leq (2D + 1)\epsilon.$$

Since $\epsilon > 0$ is arbitrary, we see that either $l(E) = 0$ or $l([0,1] - E) = 0$ (or equivalently, $\mu(E) = 0$ or $\mu([0,1] - E) = 0$).

For almost all $0 < x < 1$ we can define a sequence $a_n = a_n(x) \in \{1, 2, \dots\}$ (for $n \geq 0$) by $\frac{1}{1+a_n} < T^n(x) < \frac{1}{a_n}$. We can then write x in its *continued fraction expansion* as

$$x = \cfrac{1}{a_0 + \cfrac{1}{a_1 + \cfrac{1}{a_2 + \dots}}}.$$

Consider the function $f : (0,1) \to \mathbb{R}$ defined by $f(x) = \log n$ whenever $\frac{1}{1+n} < x < \frac{1}{n}$. Observe that

$$\int f(x) d\mu(x) = \sum_{n=1}^{\infty} \log n \frac{1}{\log 2} \int_{\frac{1}{1+n}}^{\frac{1}{n}} \frac{1}{1+x} dx$$

$$= \sum_{n=1}^{\infty} \frac{\log n}{\log 2} \log \left(\frac{(n+1)^2}{n(n+2)} \right) < +\infty.$$

By the Birkhoff theorem we have for almost all points x that

$$\frac{1}{N} \sum_{n=0}^{N-1} \log(a_n) = \frac{1}{N} \sum_{n=0}^{N-1} f(a_n) = \frac{1}{N} \sum_{n=0}^{N-1} f(T^n x) \to \int f(x) d\mu$$

as $N \to +\infty$. Taking exponentials of both sides gives that

$$\lim_{N \to +\infty} (a_1 a_2 \dots a_N)^{\frac{1}{N}} = \prod_{n=1}^{\infty} \left(\frac{(n+1)^2}{n(n+2)} \right)^{\frac{\log n}{\log 2}}$$

for almost all points x.

Application 4 (Diffeomorphisms which preserve geodesics). Let M_1 and M_2 be two finite volume smooth Riemannian surfaces whose curvatures are negative (and uniformly bounded away from zero). Let $f : M_1 \to M_2$ be a C^2 diffeomorphism.

THEOREM 10.5. *If the image of each geodesic in M_1 is a geodesic in M_2, then the diffeomorphism f is an isometry (up to a homothetic rescaling).*

We denote by $\phi_t^1 : SM_1 \to SM_1$ and $\phi_t^2 : SM_2 \to SM_2$ the associated geodesic flows on the unit tangent bundles SM_1 and SM_2 of the two manifolds. The diffeomorphism $f : M_1 \to M_2$ extends to a map $F : SM_1 \to SM_2$ defined by $F(v_x) = \frac{D_x f(v_x)}{||D_x f(v_x)||}$.

The proof is based on the following sublemmas:

SUBLEMMA 10.5.1. *Under the above hypotheses we have the following results.*

(i) *F carries orbits of ϕ^1 to orbits of ϕ^2 (but without necessarily preserving the parameterization).*

(ii) *Each flow ϕ_t^i ($i = 1, 2$) preserves the smooth Liouville measure ν_i on the sphere bundle SM_i.*

(iii) *The map F is absolutely continuous (i.e. carries sets of zero measure to sets of zero measure).*

SUBLEMMA 10.5.2.

(i) *If M_1 is a surface then for a (geodesic) arc C_1 between two points $x, y \in M_1$, the set $S_{C_1} M_1 = \cup_{x \in C_1} S_x M_1$ has relative Liouville measure $\mu_1(S_{C_1} M_1) = 2\pi d_{M_1}(x, y)$.*

(ii) *If M_1 is of dimension $n \geq 2$ then for an $(n-1)$-dimensional geodesic submanifold C_1 the set $S_{C_1} M_1 = \cup_{x \in C_1} S_x M_1$ has relative Liouville measure $\mu_2(S_{C_1} M_1) = \text{Vol}(S^{n-1})\text{Vol}_{(n-1)}(C_1)$.*

A similar conclusion holds on replacing M_1 by M_2 and $C_1 \subset M_1$ by a geodesic arc or submanifold $C_2 \subset M_2$.

By assumption, the image $C_2 := f(C_1)$ is a geodesic arc from $f(x)$ to $f(y)$. By applying sublemma 10.5.2 on M_2 we see that $\nu_2(S_{C_2} M_2) = 2\pi d_{M_2}(x, y)$.

Observe that the subsets $S_{C_1} M_1$ and $S_{C_2} M_2$ of the unit tangent bundles SM_1 and SM_2, respectively, are transverse almost everywhere to the respective flows ϕ^1 and ϕ^2. To prove Theorem 10.5 we want to show the slightly more general result that there exists a constant $C > 0$ with the property that if B_1 is a cross-section for the flow ϕ^1 (and thus $B_2 = F(B_1)$ is a cross-section for the flow ϕ^2) then $\nu_1(B_1) = C \cdot \nu_2(B_2)$. In particular, we want to apply this with $B_1 = S_{C_1} M_1$.

We shall use two simple consequences of the well-known Birkhoff ergodic theorem (presented as sublemma 10.5.3 and sublemma 10.5.4 below).

LEMMA 10.5.3 (SIZE AND RETURN TIMES). *Given a point* $x \in SM_i$ *let*

$$\pi(x,t) = Card\{0 \leq u \leq t : \phi_t^i(x) \in B_i\};$$

then for almost all $x \in SM_i$ *(with respect to Liouville measure) we have that*

$$\lim_{t \to +\infty} \frac{1}{t}\pi(x,t) = \nu_i(B_i)$$

for $i = 1,2$.

This statement and its proof essentially appear on p. 295 of [4].

Observe that since the map F preserves the orbits of the two flows, it is a conjugacy up to a change of parameterization, i.e. there exists a map $\alpha : SM_1 \times \mathbb{R} \to \mathbb{R}$ such that $F(\phi_t^1(x)) = \phi_{\alpha(x,t)}^2(Fx)$. Moreover, from the definition of F we can write $\alpha(x,t) = \int_0^t ||Df(\phi_u^1 x)||du$.

The following is a direct application of the Birkhoff ergodic theorem for flows.

LEMMA 10.5.4 (AVERAGE REPARAMETERIZATION). *For almost all points* $x \in M_1$ *we have that*

$$\lim_{t \to +\infty} \frac{\alpha(x,t)}{t} = \int ||Df(x)||d\nu_1(x).$$

PROOF OF THEOREM 10.5. For a typical point $x \in M_1$, we can write down the (increasing) sequence of times t_n such that $\phi_{t_n}^1 x \in B_1$. For the corresponding point $f(x) \in M_2$, we can write down the (increasing) sequence of times t_n' such that $\phi_{t_n'}^2 f(x) \in B_2$. Notice that by construction we have that $\pi(x,t_n) = \pi(f(x),t_n')$. By sublemma 10.5.4 we have that $t_n' \sim t_n \left(\int ||Df(x)||d\nu_1(x)\right)$ as $n \to +\infty$ (for almost all points $x \in M_1$).

For almost all $x \in M_1$ we therefore have that

$$\nu_1(B_1) = \lim_{n \to +\infty} \frac{1}{t_n}\pi(x,t_n)$$

$$= \lim_{n \to +\infty} \frac{1}{t_n}\pi(Fx,t_n')$$

$$= \left(\int ||Df(x)||d\nu_1(x)\right) \lim_{n \to +\infty} \frac{1}{t_n'}\pi(Fx,t_n')$$

$$= \left(\int ||Df(x)||d\nu_1(x)\right) \nu_2(B_2).$$

We need only choose one point x in this set of full measure to see that this identity gives us the desired identity with the choice $C = \left(\int ||Df(x)||d\nu_1(x)\right)$. ∎

10.4 The Birkhoff theorem (for invariant measures)

We begin with a definition.

DEFINITION. We let $\mathcal{I} = \{A \in \mathcal{B} : T^{-1}A = A\}$ denote the *invariant sigma-algebra*, i.e. the sigma-algebra consisting of T-invariant sets.

Notice that $E(f \circ T | \mathcal{I}) = E(f | \mathcal{I})$ for $f \in L^1(X, \mathcal{B}, \mu)$ (since $T^{-1}\mathcal{I} = \mathcal{I}$ and so $\int_A E(f \circ T | \mathcal{I}) d\mu = \int_A f \circ T d\mu = \int_{T^{-1}A} f \circ T d\mu = \int_A f d\mu$).
In the case that T is ergodic this sigma algebra is trivial i.e. $\mathcal{I} = \{X, \emptyset\}$.

THEOREM 10.6 (BIRKHOFF'S ERGODIC THEOREM). *Consider* $f \in L^1(X, \mathcal{B}, \mu)$. *If the measure μ is T-invariant then for almost all $x \in X$ we have that the averages*

$$\frac{1}{N} \sum_{n=0}^{N-1} f(T^n x) \to E(f | \mathcal{I})$$

as $N \to +\infty$, for almost all $x \in X$.

PROOF. We can assume for convenience that $E(f | \mathcal{I}) = 0$. (If this is not the case we can replace f by $f - E(f | \mathcal{I})$ on both sides of the identity.)
We need to show that if $E(f | \mathcal{I}) = 0$ then we can choose $h \in L^\infty(X, \mathcal{B}, \mu)$ with $\int |f - (hT - h)| d\mu < \delta$, i.e. $\ker (E(.|\mathcal{I})) = \mathrm{cl}(B)$. Since we are also at liberty to use sublemmas 10.5.1 and 10.5.2 as before (since their proofs did not require ergodicity) the proof will be complete.
Let $B = \{hT - h : h \in L^\infty(X, \mathcal{B}, \mu)\}$ (denoted by E in the previous proof of Birkhoff's theorem). First notice that $B \subset \ker (E(.|\mathcal{I}))$ since if $f = hT - h$ with $h \in L^\infty(X, \mathcal{B}, \mu)$ then

$$\begin{aligned}
E(hT - h | \mathcal{I}) &= E(hT | \mathcal{I}) - E(h | \mathcal{I}) \\
&= E(h | \mathcal{I})T - E(h | \mathcal{I}) \\
&= 0.
\end{aligned}$$

To show that $\ker (E(|\mathcal{I})) = \mathrm{cl}(B)$ it suffices to show that any linear functional which vanishes on B must also vanish on $\ker (E(|\mathcal{I})$. We know that any linear functional can be written in the form $f \to \int fk d\mu$, where $k \in L^\infty(X, \mathcal{B}, \mu)$. If we are assume that this functional vanishes on B then it is equivalent to $\int gk d\mu = 0$ for all $g = hT - h \in B$, i.e. $\int hTk d\mu = \int hk d\mu$, $\forall h \in L^\infty(X, \mathcal{B}, \mu)$. With the specific choice $h = k$ we get that $\int kT \cdot k d\mu = \int k^2 d\mu$ which implies that $\int (kT - k)^2 d\mu = 0$ (as in the proof of Theorem 10.2). In particular, $kT = k$.
If we assume that $f \in \ker (E(.|\mathcal{I}))$ then the image under the linear functional is

$$\int fk d\mu = \int E(fk | \mathcal{I}) d\mu = \int k E(f | \mathcal{I}) d\mu = 0$$

(by property (iii)), i.e. the functional vanishes on $\ker E(.|\mathcal{I})$. This completes the proof.

∎

10.5 Comments and references

Many simple example of applications of the ergodic theorems can be found in the appendices of [1]. A nice treatment of continued fractions and the Birkhoff ergodic theorem is contained in [2]. There exist various alternative proofs of the von Neumann ergodic theorem (cf. [3],[9]) and Birkhoff ergodic theorem (cf. [5]).

Other interesting ergodic theorems we have not considered are discussed in [6], [7] and [8, pp 101-103].

References

1. V. Arnol'd and A. Avez, *Ergodic Problems of Classical Mechanics*, Benjamin, New York, 1968.
2. P. Billingsley, *Ergodic Theory and Information*, Wiley, New York, 1965.
3. J. Bourgain, *An approach to pointwise ergodic theorems*, GAFA-seminar, 1987, Lecture Notes in Mathematics, Springer, Berlin, 1987.
4. I. Cornfeld, S. Fomin and Y. Sinai, *Ergodic Theory*, Springer, Berlin, 1982.
5. Y. Katznelson and B. Weiss, *A simple proof of some ergodic theorems*, Israel J. Math. **42** (1982), 291-296.
6. U. Krengel, *Ergodic Theorems*, de Gruyter, Berlin, 1985.
7. W. Parry, *Topics in Ergodic Theory*, C.U.P., Cambridge, 1986.
8. K. Petersen, *Ergodic Theory*, C.U.P., Cambridge, 1983.
9. R. Zimmer, *Essential Results of Functional Analysis*, Chicago University Press, Chicago, 1990.

MIXING PROPERTIES

We now want to consider two stronger properties than ergodicity. These are weak mixing and strong mixing which are important from the statistical point of view, as we shall see in the next chapter.

11.1 Weak mixing

DEFINITION. Let $T : X \to X$ be a measure preserving map on a probability space $(X, \mathcal{B}, \mu)$; then we call T *weak-mixing* if for any $A, B \in \mathcal{B}$ we have that

$$\frac{1}{N} \sum_{n=0}^{N-1} |\mu(T^{-n} A \cap B) - \mu(A)\mu(B)| \to 0$$

as $N \to +\infty$.

We have the following equivalent characterization.

LEMMA 11.1. *The following are equivalent.*

(i) *T is weak-mixing;*
(ii) *for $f, g \in L^2(X, \mathcal{B}, \mu)$ we have that*

$$\frac{1}{N} \sum_{n=0}^{N-1} | \int f T^n g d\mu - \int f d\mu \int g d\mu| \to 0$$

as $N \to +\infty$.

PROOF. For "(ii) implies (i)" we need only make the choices $f = \chi_A$ and $g = \chi_B$. For "(i) implies (ii)" we can use an argument of approximation by step functions (finite linear combinations of characteristic functions). ∎

The following lemma shows that weak-mixing is a stronger property than ergodicity.

LEMMA 11.2. *If a transformation* $T : X \to X$ *on a probability space* $(X, \mathcal{B}, \mu)$ *is weak-mixing then it is necessarily ergodic.*

PROOF. If T is weak-mixing then by definition we have that for any $A, B \in \mathcal{B}$

$$\frac{1}{N} \sum_{n=0}^{N-1} |\mu(T^{-n}A \cap B) - \mu(A)\mu(B)| \to 0$$

as $N \to +\infty$. By the triangle inequality we have that

$$
\begin{aligned}
&|\frac{1}{N} \sum_{n=0}^{N-1} \mu(T^{-n}A \cap B) - \mu(A)\mu(B)| \\
&\leq \frac{1}{N} \sum_{n=0}^{N-1} |\mu(T^{-n}A \cap B) - \mu(A)\mu(B)| \\
&\to 0.
\end{aligned}
\tag{11.1}
$$

If we assume (for a contradiction) that T were not ergodic then there would exist a T-invariant set $E \in \mathcal{B}$ with $T^{-1}E = E$ with $0 < \mu(E) < 1$. If we take $A = E$ and $B = X - E$ in (11.1) then since $\mu(T^{-n}E \cap (X - E)) = \mu(E \cap (X - E)) = 0$, for all $n \geq 0$, we deduce that $\mu(E)cdot\mu(X - E) = 0$ giving the required contradiction. Thus T is ergodic. ∎

The converse is not true: there exist examples of transformations which are ergodic but not weak-mixing, as the following simple example shows.

EXAMPLE (ERGODIC, NOT WEAK-MIXING). Let $X = \mathbb{R}/\mathbb{Z}$, let $\mathcal{B}$ be the Borel sigma-algebra, and let μ be the Haar-Lebesgue measure. For any irrational number $a \in \mathbb{R}$ the transformation $T : X \to X$ defined by $T(x) = x + a \pmod 1$ is known to be ergodic. We can see that it is not weak-mixing by choosing $A = B = [0, \frac{1}{2}]$, and then $\mu(A)\mu(B) = \frac{1}{4}$. Since the sequence $na + \mathbb{Z}$ is uniformly distributed we know that the proportion of the terms in the sub-sequence n_i for which $n_i a \in [0, \frac{1}{100}] \pmod 1$ is $\frac{1}{100}$. For these terms we have that $\mu(T^{n_i}A \cap B) - \mu(A)\mu(B) \geq \frac{49}{100} - \frac{1}{4} = \frac{24}{100}$ which means that

$$\liminf_{N \to +\infty} \frac{1}{N} \sum_{n=0}^{N-1} |\mu(T^{n_i}A \cap B) - \mu(A)\mu(B)| \geq \frac{1}{100} \cdot \frac{24}{100} > 0.$$

11.2 A density one convergence characterization of weak mixing

Using the previous lemma on sequences there is also a characterization for weak mixing which is closer to that of strong mixing. We say that a sequence $\{n_i\}_{i \in \mathbb{N}}$ of natural numbers has density one if

$$\lim_{n \to +\infty} \frac{1}{n} \mathrm{Card}\{n_i \in [0, 1, \ldots, n-1] : i \in \mathbb{N}\} = 1.$$

PROPOSITION 11.3. *The transformation T is weak-mixing if there exists a sequence $\{n_i\}_{i \in \mathbb{N}}$ of density one such that $\mu(T^{-n_i} A \cap B) \to \mu(A)\mu(B)$ as $n_i \to +\infty$.*

The proof requires only the following simple lemma on sequences.

LEMMA 11.4. *The following are equivalent for a bounded sequence of real numbers $\{a_n\}$.*

(1) $\frac{1}{n} \sum_{k=0}^{n-1} a_k \to 0$ *as $n \to +\infty$; and*

(2) $\lim_{k \to +\infty} a_{n_k} = 0$ *for some sub-sequence $\{n_k\} \subset \mathbb{N}$ of density one (i.e. $\lim_{n \to +\infty} \frac{1}{n} \mathrm{Card}\{n_i \in [0, 1, \dots, n-1]: i \in \mathbb{N}\} = 1$).*

PROOF. (2) $\implies$ (1): Let J be a sequence of density one. Given $\epsilon > 0$ we can choose N such that for $n \geq N$ we have that

$$Card\{n_i \in [0, 1, \dots, n-1]\} \geq n(1 - \epsilon)$$

and for $n_i \geq N$ we have $|a_{n_i}| \leq \epsilon$. In particular,

$$\left| \frac{1}{n} \sum_{k=0}^{n-1} a_k \right| = \frac{1}{n} \left(\sum_{k=0}^{N} |a_k| + \sum_{\substack{k \in \{n_i\} \\ N < k \leq n-1}} |a_k| + \sum_{\substack{k \notin \{n_i\} \\ N < k \leq n-1}} |a_k| \right)$$

$$\leq \frac{1}{n} \sum_{k=0}^{N} |a_k| + \epsilon \left((1 - \epsilon) + \sup\{|a_i|\} \right).$$

We can choose n sufficiently large that $\frac{1}{n} \sum_{k=0}^{N} |a_k| < \epsilon$. Since $\epsilon > 0$ can be chosen arbitrarily small we have that $\frac{1}{n} \sum_{k=0}^{n-1} a_k \to 0$.

(1) $\implies$ (2): Assume that $\frac{1}{n} \sum_{k=0}^{n-1} a_k \to 0$. For each $m \geq 1$ we define

$$J_m = \{n \in \mathbb{N} : a_n \leq \frac{1}{m}\}$$

and observe that this has density one, since

$$\frac{1}{m} \left(\frac{1}{n} \sum_{k=0}^{n-1} \chi_{\mathbb{N} - J_m}(k) \right) \leq \frac{1}{n} \sum_{k=0}^{n-1} a_k \to 0$$

as $n \to +\infty$. Thus for each $m \geq 1$ we can choose n_m such that

$$\frac{1}{n} \left(\sum_{k=0}^{n-1} \chi_{\mathbb{N} - J_m}(k) \right) \leq \frac{1}{m}$$

for $n \geq n_m$. We then define

$$J = \cup_{k=1}^{\infty} J_k \cap [N_k, \dots, N_{k+1}]$$

and it is easy to see that J has density one and $\lim_{n \to \infty : n \in J} a_n = 0$. ∎

The following corollary is quite useful.

COROLLARY 11.4.1. $\frac{1}{n}\sum_{k=0}^{n-1} a_k \to 0$ as $n \to +\infty$ if and only if $\frac{1}{n}\sum_{k=0}^{n-1} a_k^2$ $\to 0$ as $n \to +\infty$.

11.3 A generalization of the von Neumann ergodic theorem

We want to present the following interesting generalization of the Von Neumann ergodic theorem (Theorem 10.1) for weak-mixing transformations. It will only be used in chapter 16 and is not required for the rest of this chapter.

THEOREM 11.5. *Assume that* $f_1, \ldots, f_k \in L^\infty(X, \mathcal{B}, \mu)$. *If* $T : X \to X$ *is weak-mixing then*

$$\frac{1}{N}\sum_{n=1}^{N} f_1(T^n x) f_2(T^{2n} x) \ldots f_k(T^{kn} x) \to \int f_1 d\mu . \int f_2 d\mu \ldots \int f_k d\mu$$

(in the L^2 topology) as $N \to +\infty$.

PROOF. The proof is by induction. When $k = 1$, this is precisely the Von Neumann ergodic theorem (Theorem 10.1).

Assume that the result has been established for $k - 1$ functions. We may assume without loss of generality that $\int f_k d\mu = 0$ (otherwise we need only replace f_k by $f_k - \int f_k d\mu$). Thus it suffices to show that

$$\int |\frac{1}{N}\sum_{n=1}^{N} f_1(T^n x) f_2(T^{2n} x) \ldots f_k(T^{kn} x)|^2 d\mu(x) \to 0 \text{ as } N \to +\infty.$$

For any $1 \le m \le N$ we can now bound

$$\int |\frac{1}{N}\sum_{n=1}^{N} f_1(T^n x) f_2(T^{2n} x) \ldots f_k(T^{kn} x)|^2 d\mu(x)$$

$$\le \int \left| \frac{1}{N}\sum_{n=1}^{N} \left(\frac{1}{m}\sum_{j=0}^{m-1} f_1(T^{n+j} x) f_2(T^{2(n+j)} x) \ldots f_k(T^{k(n+j)} x) \right) \right|^2 d\mu(x)$$

$$+ \left(\frac{2m}{N} + \frac{m^2}{N^2} \right) \left(\max_{1 \le i \le k} |f_i|_\infty \right)^2$$

$$\le \frac{1}{N}\sum_{n=1}^{N} \left(\int |\frac{1}{m}\sum_{j=0}^{m-1} f_1(T^{n+j} x) f_2(T^{2(n+j)} x) \ldots f_k(T^{k(n+j)} x)|^2 d\mu(x) \right)$$

$$+ \left(\frac{2m}{N} + \frac{m^2}{N^2} \right) \left(\max_{1 \le i \le k} |f_i|_\infty \right)^2 .$$

$$(11.2)$$

(The first inequality comes from the observation that

$$\frac{1}{N}\sum_{n=1}^{N} a_n \leq \frac{1}{N}\sum_{n=1}^{N-(m-1)}\left(\frac{1}{m}\sum_{j=0}^{m-1} a_{n+j}\right) + \frac{m}{N}\left(\max_{1\leq i\leq m}|a_i| + \max_{N-m\leq i\leq N}|a_i|\right)$$

for any real numbers $a_1,\dots,a_N$. The second inequality comes from the observation that $\left(\frac{1}{N}\sum_{n=1}^{N} b_n\right)^2 \leq \frac{1}{N}\sum_{n=1}^{N}|b_n|^2$ for any real numbers $b_1,\dots,$ b_N.)   We next observe that

$$\int |\sum_{j=0}^{m-1} f_1(T^{n+j}x)f_2(T^{2(n+j)}x)\dots f_k(T^{k(n+j)}x)|^2 d\mu(x)$$

$$= \sum_{i=0}^{m-1}\sum_{j=0}^{m-1}\int\left(\prod_{l=1}^{k} f_l(T^{l(n+i)}x)\right)\left(\prod_{l=1}^{k} f_l(T^{l(n+j)}x)\right)d\mu(x) \qquad (11.3)$$

$$= \sum_{i=0}^{m-1}\sum_{j=0}^{m-1}\int\left(\prod_{l=1}^{k}\left(f_l.f_l\circ T^{l(j-i)}\right)(T^{l(n+i)}x)\right)d\mu(x).$$

By the inductive hypothesis we know that for each $0\leq i,j\leq m-1$,

$$\frac{1}{N}\sum_{n=1}^{N}\prod_{l=2}^{k}\left(f_l\cdot f_l\circ T^{l(j-i)}\right)(T^{l(n+i)}x) \to \prod_{l=2}^{k}\int f_l\cdot f_l\circ T^{l(j-i)}d\mu$$

as $N\to+\infty$ (in the L^2 topology) and so

$$\frac{1}{N}\sum_{n=1}^{N}\int\prod_{l=1}^{k}\left(f_l.f_l\circ T^{l(j-i)}\right)(T^{l(n+i)}x)d\mu(x)$$

$$= \frac{1}{N}\sum_{n=1}^{N}\int\left(f_1\cdot f_1\circ T^{(j-i)}\right)(x)\left(\prod_{l=2}^{k}\left(f_l\cdot f_l\circ T^{l(j-i)}\right)(T^{l(n+i)}x)\right)d\mu(x)$$

$$\to \prod_{l=1}^{k}\int f_l\cdot f_l\circ T^{l(j-i)}d\mu.$$

$$(11.4)$$

Comparing (11.1), (11.2) and (11.3) we see that

$$\limsup_{N\to+\infty}\int|\frac{1}{N}\sum_{n=1}^{N} f_1(T^n x)f_2(T^{2n}x)\dots f_k(T^{kn}x)|^2 d\mu(x)$$

$$(11.5)$$

$$\leq \frac{1}{m^2}\sum_{i=0}^{m-1}\sum_{j=0}^{m-1}\left(\prod_{l=1}^{k}\int f_l\cdot f_l\circ T^{l(j-i)}d\mu(x)\right).$$

Finally, since T is weak-mixing we know that $\int f_l\cdot f_l\circ T^{lr_n}d\mu(x)\to\int f_l d\mu(x)$ $=0$ where $r_n\to+\infty$ through a set of density one. Thus for sufficiently large m the expression in (11.4) can be made arbitrarily small.

∎

11.4 The spectral viewpoint

Consider a measure preserving transformation $T : X \to X$ on a probability space $(X, \mathcal{B}, \mu)$. Consider the Hilbert space $H = L^2(X, \mathcal{B}, \mu)$ of square integrable functions with the inner product $\langle f, g \rangle = \int f \bar{g} d\mu$. We can associate to T an operator $U_T : H \to H$ defined by $(U_T f)(x) = f(Tx)$ whenever $f \in H$. It is easy to see the following.

LEMMA 11.6. $U_T : H \to H$ is an isometry. i.e. $\|U_T f\| = \|f\|$.

We recall a few elementary observations about operators $U : H \to H$ on a Hilbert space H. We call a linear operator U an *isometry* if for every pair of vectors $x, y \in H$ we have that $\langle Ux, Uy \rangle = \langle x, y \rangle$. We shall only be interested in isometries.

An eigenvalue for $U : H \to H$ is a complex number $\alpha \in \mathbb{C}$ for which there exists a (non-zero) vector $x \in H$ (called the eigenvector) such that $Ux = \alpha x$.

LEMMA 11.7. *Eigenvalues of isometries must be complex numbers of modulus unity.*

PROOF. Clearly, if $Ux = \alpha x$ then $\langle Ux, Ux \rangle = \alpha \bar{\alpha} \langle x, x \rangle$. But since U is an isometry we have that $\langle Ux, Uy \rangle = \langle x, y \rangle$ and so $\langle x, x \rangle = |\alpha|^2 \langle x, x \rangle$ i.e. $|\alpha| = 1$, as claimed.

An important aspect of the spectrum of the operator is the variety (or lack of it) of eigenvectors. Two extreme cases are the following.

DEFINITION. The operator $U : H \to H$ has *continuous spectrum* if there are no eigenvectors. The operator $U : H \to H$ has *pure point spectrum* if H is the closure of the linear span of the eigenvectors.

REMARK. Between these two extreme cases we have the possibility of having *mixed spectrum*. We can let $V \subset H$ denote the subspace spanned by the eigenvectors. We let $V^\perp \subset H$ denote the orthogonal subspace to V. From the definitions we see that $U : V \to V$ then has pure point spectrum and $U : V^\perp \to V^\perp$ has continuous spectrum.

We now recall one of the basic theorems in spectral theory.

DEFINITION. A sequence $r_n \in \mathbb{C}$, $n \geq 0$, is called *positive definite* if for each $N \geq 1$ and each sequence $a_0, \ldots, a_N$ we always have that

$$\sum_{n,m=0}^{N} r_{n-m} a_n \bar{a}_m \geq 0.$$

BOCHNER-HERGLOTZ SPECTRAL THEOREM. *If r_n, $n \geq 0$, is positive definite then there is a unique finite Borel measure on $\mathbb{R}/\mathbb{Z}$ such that $r_n = \int_0^1 e^{2\pi i n t} d\mu(t)$.*

(The proof can be found in the appendix to [1].)

APPLICATION TO ISOMETRIES. The Bochner-Herglotz theorem is particularly well suited to isometries $U : H \to H$. Fix $x \in H$ and then set $r_n = \langle U^n x, x \rangle$ and $r_{-n} = \langle x, U^n x \rangle$ and observe that

$$\sum_{n,m=0}^{N} r_{n-m} a_n \bar{a}_m = \langle \sum_{n=0}^{N} a_n U^n x, \sum_{n=0}^{N} a_n U^n x \rangle \geq 0.$$

The measure μ on the unit circle $\mathbb{R}/\mathbb{Z}$ is called the *spectral measure*.

The choice of point $x \in H$ affects the resulting spectral measure μ. If $x \in V$ (i.e. x is in the closure of the span of the eigenvectors) then the associated measure is singular with respect to the Haar-Lebesgue measure on the circle $\mathbb{R}/\mathbb{Z}$. If $x \in V^\perp$ then the associated measure is absolutely continuous with respect to the Haar-Lebesgue measure on the circle $\mathbb{R}/\mathbb{Z}$.

EXAMPLE. Consider an irrational rotation $T : X \to X$ on the unit circle $X = \mathbb{R}/\mathbb{Z}$ defined by $T(x + \mathbb{Z}) = (x + a + \mathbb{Z})$. The functions $e_n(x) = e^{2\pi i n x}$ are eigenfunctions for the operator $U : L^2(X, \mathcal{B}, \mu) \to L^2(X, \mathcal{B}, \mu)$ since

$$U e_n(x) = e^{2\pi i n(x+a)} = e^{2\pi i n a} e_n(x).$$

Since the family $e_n, n \in \mathbb{Z}$, spans the space the transformation T has pure point spectrum.

In applying this to ergodic theory, we consider a measure preserving transformation $T : X \to X$ on the Hilbert space $H = L^2(X, \mathcal{B}, \mu)$ with inner product $< f, g >= \int f \bar{g} d\mu$.

REMARK. The following lemma is also a standard result from spectral theory (although we won't require it).

RIEMANN-LEBESGUE LEMMA. *If the spectral measure μ on $\mathbb{R}/\mathbb{Z}$ for the operator $U : H \to H$ (and a point $x \in H$) is absolutely continuous then $< U^n x, x > \to 0$ as $n \to +\infty$.*

11.5 Spectral characterization of weak mixing

The Hilbert space $H = L^2(X, \mathcal{B}, \mu)$ has the obvious one-dimensional subspace consisting of constant functions and denoted by $\mathbb{C}$. We let $\mathbb{C}^\perp$ denote the orthonormal (co-dimension one) subspace. For a measure preserving transformation $T : X \to X$ the associated isometry $U : H \to H$ preserves both $\mathbb{C}$ and $\mathbb{C}^\perp$.

PROPOSITION 11.8. *Let $T : X \to X$ be a measure preserving transformation on the probability space $(X, \mathcal{B}, \mu)$. The following conditions are equivalent:*

(1) *for the map $U : C^\perp \to C^\perp$ has continuous spectrum;*
(2) *T is weak-mixing;*
(3) *the measure preserving transformation $T \times T : X \times X \to X \times X$ (on $X \times X$ with the product measure $\mu \times \mu$) defined by $(T \times T)(x, y) = (Tx, Ty)$ is weak-mixing (and thus ergodic).*

PROOF.

(1) $\implies$ (2): Assume that $U : C^\perp \to C^\perp$ has continuous spectrum. Choose any vector $f \in C^\perp$; then we estimate that

$$\frac{1}{N} \sum_{n=0}^{N-1} |\int f \circ T^n \bar{f} d\mu|^2$$

$$= \frac{1}{N} \sum_{n=0}^{N-1} \left(\int f \circ T^n \bar{f} d\mu \right) \left(\int \bar{f} \circ T^n f d\mu \right)$$

$$= \frac{1}{N} \sum_{n=0}^{N-1} \left(\int_0^1 e^{2\pi i n t} d\bar{\mu}(t) \right) \left(\int_0^1 e^{-2\pi i n s} d\bar{\mu}(s) \right)$$

$$= \frac{1}{N} \sum_{n=0}^{N-1} \int_0^1 \int_0^1 e^{2\pi i n(t-s)} d(\bar{\mu} \times \bar{\mu})(t, s)$$

$$= \int_0^1 \int_0^1 \frac{1}{N} \left(\frac{e^{2\pi i N(t-s)} - 1}{e^{2\pi i (t-s)} - 1} \right) d(\bar{\mu} \times \bar{\mu})(t, s).$$

Observe that since μ has continuous spectrum the product measure $\bar{\mu} \times \bar{\mu}$ gives the diagonal $\{(t, t) : t \in \mathbb{R}/\mathbb{Z}\}$ measure zero. In particular, the last integrand is finite almost everywhere. Since

$$\frac{1}{N} \frac{e^{2\pi i N(t-s)} - 1}{e^{2\pi i (t-s)} - 1} \to 0$$

for almost all (s, t) and it is dominated by the constant function 1, we see by the Lebesgue dominated convergence theorem that

$$\frac{1}{N} \sum_{n=0}^{N-1} | \int f \circ T^n \bar{f} |^2 \leq \int_0^1 \int_0^1 \left(\frac{1}{N} \frac{e^{2\pi i N(t-s)} - 1}{e^{2\pi i (t-s)} - 1} \right) d(\bar{\mu} \times \bar{\mu})(t, s) \to 0$$

as $N \to +\infty$. Finally, we observe that

$$\frac{1}{N} \sum_{n=0}^{N-1} | \int f \circ T^n \bar{f} d\mu |^2 \to 0$$

implies

$$\frac{1}{N} \sum_{n=0}^{N-1} \int f \circ T^n \bar{f} \to 0$$

(by Corollary 11.4.1).

(2) $\implies$ (3):   Consider sets $E = \sup_i(A_i \times B_i), F = \sup_j(C_j \times D_i)$ (for finite disjoint unions of product sets A_i, B_i, C_i and D_i). Since T is weak-mixing we know that

$$\mu\left(T^{-n_k} A_i \cap C_j\right) \to \mu(A_i)\mu(C_i) \tag{11.6}$$

and

$$\mu\left(T^{-n_k} B_i \cap D_j\right) \to \mu(B_i)\mu(D_i) \tag{11.7}$$

for sequences $n_k \to +\infty$ of density one (and without loss of generality we can assume that we have the same sequence in each case).

To show $T \times T$ is weak-mixing we want to show that

$$(\mu \times \mu)\left((T \times T)^{-n_i} E \cap F\right) \to (\mu \times \mu)(E)(\mu \times \mu)(F)$$

for a sequence $n_k \to +\infty$ of unit density. This follows from (11.5) and (11.6) since

$$\begin{aligned}
(\mu \times \mu)&\left((T \times T)^{-n_i}(A_i \times B_i) \cap (C_j \times D_j)\right) \\
&= \mu\left(T^{-n_i} A_i \cap C_j\right) \cdot \mu\left(T^{-n_i} B_i \cap D_j\right) \\
&\to \mu(A_i)\mu(C_i)\mu(B_i)\mu(D_i) \\
&= \mu\left(T^{-n_i} A_i \cap C_j\right) \cdot \mu\left(T^{-n_i} B_i \cap D_j\right)
\end{aligned}$$

for a sequence $n_k \to +\infty$ of unit density.

(3) $\implies$ (1):   Assume for a contradiction that there is a *non-constant* eigenfunction $f \in H$ for $T : X \to X$, i.e. $Uf = \alpha f$. We can then define a function $F : X \times X \to \mathbb{C}$ by $F(x, y) = f(x)\bar{f}(y)$. Observe that

$$F(Tx, Ty) = f(Tx)\bar{f}(Ty) = \alpha \bar{\alpha} f(x) \bar{f}(y) = F(x, y).$$

But then $F(x, y)$ is a $(T \times T)$-invariant function which is non-constant. This contradicts $T \times T$ being ergodic (and therefore being weak-mixing).

11.6 Strong mixing

We now turn to another notion of "mixing".

DEFINITION. Let $T : X \to X$ be a measure preserving transformation on a probability space $(X, \mathcal{B}, \mu)$; then we call T *strong mixing* if for any $A, B \in \mathcal{B}$ we have that

$$\mu(T^{-n} A \cap B) \to \mu(A)\mu(B)$$

as $n \to +\infty$. We have the following equivalent characterization.

LEMMA 11.9. *The following are equivalent:*

(i) T *is strong-mixing;*
(ii) *for* $f, g \in L^2(X, \mathcal{B}, \mu)$, $\int f \circ T^n g d\mu \to \int f d\mu \int g d\mu$ *as* $n \to +\infty$.

PROOF. For "(ii) $\implies$ (i)" we need only make the choices $f = \chi_A$ and $g = \chi_B$. For "(i) $\implies$ (ii)" we can use an argument of approximation by step functions (finite linear combinations of characteristic functions). ∎

The following lemma states the obvious fact that strong mixing is a stronger property than weak mixing.

LEMMA 11.10. *If* T *is a strong-mixing transformation on a probability space* $(X, \mathcal{B}, \mu)$ *then it is necessarily weak-mixing (and thus also ergodic).*

PROOF. This is immediate from the definitions. ∎

EXAMPLE (MARKOV MEASURES AND SHIFTS). Recall that a subshift of finite type $T : X \to X$ is defined on a space

$$X = \{x \in \prod_{n \in \mathbb{Z}} \{1, \dots, k\} : A(x_n, x_{n+1}) = 1, n \in \mathbb{Z}\}$$

for some $k \times k$ matrix A with entries either zero or unity. We define $T(x_n) = (x_{n+1})$ (i.e. all terms in the infinite sequences are shifted one place to the left). We shall assume in addition that A is aperiodic, i.e. there exists $n \geq 1$ such that for each $1 \leq i, j \leq k$ we have that $A^n(i, j) = 1$.

Let P denote a $k \times k$ stochastic matrix with entries $P(i, j) = 0$ iff $A(i, j) = 0$, and let p be its left eigenvector. Recall that we define the associated Markov measure by

$$\mu[i_0, \dots, i_{l-1}] = p(i_0) P(i_0, i_1) \dots P(i_{l-2}, i_{l-1}).$$

Let $C = [i_0, \ldots, i_{l-1}]$ and $D = [j_0, \ldots, j_{l-1}]$ be two cylinder sets. Observe that

$$C \cap T^{-(n+l)}D = \cup_{x_l, \ldots, x_{n+l-1}}[i_0, \ldots, i_{l-1}, x_l, \ldots, x_{n+l-1}, j_0, \ldots, j_{l-1}]$$

and thus

$$\mu(C \cap T^{-(n+l)}D)$$
$$= \sum_{x_l, \ldots, x_{n+l-1}} p(i_0)P(i_0, i_1) \ldots P(i_{l-1}, x_l) \ldots P(x_{n+l-1}, j_0) \ldots P(j_{l-2}, j_{l-1})$$
$$= \mu(C)\mu(D)\left(\frac{1}{p(j_0)}\sum_{x_l, \ldots, x_{n+l-1}} P(i_{l-1}, x_l) \ldots P(x_{n+l-1}, j_0)\right)$$
$$= \mu(C)\mu(D)\frac{P^n(i_{l-1}, j_0)}{p(j_0)}.$$

However, we know that $P^n(i_{l-1}, j_0) \to p(j_0)$ as $n \to +\infty$ (by writing P in terms of its eigenvectors) and so we know that

$$\mu(C \cap T^{-(n+l)}D) \to \mu(C)\mu(D)$$

as $n \to +\infty$. For arbitrary sets $A, B \in \mathcal{B}$ we can cover them by unions of disjoint cylinders $A \subset \cup_{i=1}^n C_i$ and $B \subset \cup_{j=1}^m D_i$ such that $\mu((\cup_{i=1}^n C_i - C) < \epsilon$ and $\mu((\cup_{i=1}^m D_i - D) < \epsilon$ and then by approximation we see that $\mu(C \cap T^{-(n+l)}D) \to \mu(C)\mu(D)$ as $n \to +\infty$.

11.7 Comments and reference

We have given only the briefest introduction to the spectral theory associated with measure preserving transformations. A particularly nice introduction is contained in the appendix to [1].

Reference

1. W. Parry, *Topics in Ergodic Theory*, C.U.P., Cambridge, 1981.

CHAPTER 12

STATISTICAL PROPERTIES IN ERGODIC THEORY

12.1 Exact endomorphisms

DEFINITION. We call a measure preserving transformation $T : X \to X$ on a probability space $(X, \mathcal{B}, \mu)$ an *exact endomorphism* if $\cap_{n=0}^{\infty} T^{-n}\mathcal{B} = \{X, \emptyset\}$ up to a set of zero measure (i.e. if $B \in T^{-n}\mathcal{B}$, for every $n \geq 0$, then $\mu(B) = 0$ or $\mu(B) = 1$).

PROPOSITION 12.1. $T : X \to X$ *is exact if for any positive measure set* A *with* $T^n A \in \mathcal{B}(n \geq 0), \mu(T^n(A)) \to 1$ *as* $n \to +\infty$.

It is easy to see that this sufficient condition for exactness is also necessary [2] (although we will not need this here).

PROOF. First we remark that T is exact if every measurable set A satisfying for arbitrary n the relationship $A = T^{-n}(T^n A)$ is of either measure zero or measure 1. For such a set A, it is clear that $\mu(A) = 1$ if $\mu(A) > 0$, as $\mu(T^n A) = \mu(A)$ and so $\lim_{n \to \infty} \mu(T^n A) = \mu(A) = 1$ if $\mu(A) > 0$. ∎

PROPOSITION 12.2. *If* T *is exact then it is strong-mixing.*

PROOF. Consider the sub-sigma-algebras $\mathcal{B} \supset T^{-1}\mathcal{B} \supset T^{-2}\mathcal{B} \supset \ldots \supset \{X, \emptyset\}$. We can associate the nested subspaces $L^2(\mathcal{B}) \supset L^2(T^{-1}\mathcal{B}) \supset L^2(T^{-2}\mathcal{B}) \supset \ldots \supset \mathbb{C}$ and for each $n \neq 0$ we can choose an orthonormal basis $\{k_i\}_{i=0}^{\infty}$ for $L^2(T^n\mathcal{B}) \ominus L^2(T^{n+1}\mathcal{B})$. It follows that $\{k_i \circ T^n\}_{i=0 \, n=0}^{\infty \, \infty}$ is an orthonormal basis for $L^2(X, B, \mu)$. Two functions $f, g \in L^2(X, B, \mu) \ominus \mathbb{R}$ can be written in the form

$$\begin{cases} f = \sum_{n=0}^{\infty} \sum_i a_{n,i} k_i \circ T^n + \left(\int f d\mu \right), \\ g = \sum_{n=0}^{\infty} \sum_i b_{n,i} k_i \circ T^n + \left(\int g d\mu \right), \end{cases}$$

where $a_{n,i}, b_{n,i} \in \mathbb{R}$. In particular,

$$\int f \circ T^N g d\mu = \sum_{n=0}^{\infty} \sum_i a_{n,i} b_{n+N,i} + \int f d\mu \int g d\mu \to \int f d\mu \int g d\mu$$

as $N \to +\infty$, i.e. T is strong-mixing. ∎

EXAMPLE 1 (ONE-SIDED MARKOV SHIFTS). We can modify the definition of the Markov shift and define

$$X_A^+ = \{x \in \prod_{n \in \mathbb{N}^+} \{0, \dots, k-1\} \, : \, A(x_n, x_{n+1}) = 1, n \in \mathbb{Z}^+\}$$

and $\sigma : X_A^+ \to X_A^+$ by $(\sigma x)_n = x_{n+1}$. For the stochastic matrix P (with entries $P(i,j) = 0$ iff $A(i,j) = 0$) letting p be its left eigenvector we define the measure on a cylinder

$$[i_0, \dots, i_{l-1}] = \{x \in X_A^+ \, : \, x_j = i_j, 0 \le j \le l-1\},$$

$$\mu[i_0, \dots, i_{l-1}] = p(i_0)P(i_0, i_1) \dots P(i_{l-2}, i_{l-1}).$$

The argument for the (two sided) Markov shift still applies and we see that T is strong-mixing; moreover, $\forall \epsilon > 0, \forall$ cylinders C, $\exists N > 0$ such that $\forall n \ge N$ and any cylinder D we have $|\mu(C \cap T^{-n}D) - \mu(C)\mu(D)| \le \epsilon\mu(C)\mu(D)$. By approximating an arbitrary set $B \in \mathcal{B}$ by a cylinder D we see that the same result holds on replacing D by B.

Assume that $E \in \cap_{n=0}^\infty T^{-n}\mathcal{B}$ and write $E = T^{-n}E_n$. For any cylinder C we see from the above observations that

$$\mu(C \cap E) = \mu(C \cap T^{-n}E_n) \ge (1 - \epsilon)\mu(E_n)\mu(C) = (1 - \epsilon)\mu(E)\mu(C);$$

since $\epsilon > 0$ is arbitrary we see that $\mu(C \cap E) \ge \mu(E)\mu(C)$ for all cylinders C. By approximation by disjoint unions of cylinders we can replace this by $\mu(B \cap E) \ge \mu(E)\mu(B)$, $\forall B \in \mathcal{B}$. If we take $B = X - E$ we see that $\mu(E)\mu(X - E) = 0$. This completes the proof that T is exact.

12.2 Statistical properties of piecewise expanding Markov maps

Consider a piecewise expanding C^2 surjective Markov map $T : I \to I$ for which there exists $\beta > 1$ with $\inf_{x \in I} |T'(x)| \ge \beta$. We can define an operator $\mathcal{L} : L^1(I) \to L^1(I)$ as follows.

DEFINITION. Given $f \in L^1(I)$ we define the *Perron-Frobenius operator* by

$$\mathcal{L}f(x) = \sum_{y \in T^{-1}x} \frac{f(y)}{|T'(y)|} \left(= \sum_{i=1}^k f(\psi_i x)|\psi_i'(x)|\chi_{TI_i}(x) \right)$$

(where ψ_i denotes the inverse of $T|I_i$).

LEMMA 12.3. *For any $f \in L^1(I)$ satisfying $(\mathcal{L}f)(x) = f(x)$ the measure μ defined by $f = \frac{d\mu}{dx}$ is T-invariant.*

PROOF. This follows from the change of variables formula since we have
$\mu(T^{-1}A) = \int_A f(x)dx = \sum_{i=1}^k \int_{TI_i \cap A} |\psi_i'(x)| f \circ \psi_i(x)dx = \int_A \mathcal{L}f(x)dx = \mu(A)$.

∎

We have the following result.

PROPOSITION 12.4 (SMOOTH INVARIANT MEASURES FOR PIECEWISE EX-PANDING MARKOV MAPS). *There exists an invariant probability measure μ which is absolutely continuous with respect to the (normalized) Haar-Lebesgue measure λ (i.e. there exists $f \in L^1(I)$ such that $\mu(B) = \int_B f(x)d\lambda(x)$ for every Borel set $B \in \mathcal{B}$).*

PROOF. By Lemma 12.3, to construct μ it suffices to find such a function f satisfying $\mathcal{L}f = f$. We first choose a point $x \in I$ and for any $n \geq 1$ we look at the families $T^{-n}x$ of all n-iterate pre-images of x.

It is easy to see from the chain rule that

$$\mathcal{L}^n 1(x) = \sum_{y \in T^{-1}x} \frac{\mathcal{L}^{n-1}1(y)}{|T'(y)|} = \sum_{y \in T^{-n}x} \frac{1}{|T^{n\prime}(y)|}.$$

We denote the inverse of $T^n| \cap_{j=0}^{n-1} T^{-j}I_{i_{j+1}}$ by $\psi_{i_1 \ldots i_n}$. Let $\mathcal{V}$ be the partition generated by $\{T(I_i) : 1 \leq i \leq k\}$. Then for $x, x' \in V \in \mathcal{V}$ we can compare

$$|\mathcal{L}^n 1(x) - \mathcal{L}^n 1(x')| = |\sum_{y \in T^{-n}x} \frac{1}{|T^{n\prime}(y)|} - \sum_{y' \in T^{-n}x'} \frac{1}{|T^{n\prime}(y')|}|$$

$$= \sum_{i_1,\ldots,i_n} \left| |\psi_{i_1 \ldots i_n}'(x)| - |\psi_{i_1 \ldots i_n}'(x')| \right| \chi_{T^n I_{i_1 \ldots i_n}}(x)$$

where $I_{i_1 \ldots i_n} = \cap_{j=0}^{n-1} T^{-j} I_{i_{j+1}}$. Observe that

$$\log \left| \frac{\psi_{i_1 \ldots i_n}'(x')}{\psi_{i_1 \ldots i_n}'(x)} \right| = \sum_{j=1}^n \log \left| \frac{\psi_{i_j}'(\psi_{i_{j+1} \ldots i_n}x')}{\psi_{i_j}'(\psi_{i_{j+1} \ldots i_n}x)} \right|$$

$$= \sum_{j=1}^n \log \left| \frac{T'(\psi_{i_j \ldots i_n}x')}{T'(\psi_{i_j \ldots i_n}x)} \right|$$

$$\leq \sum_{j=1}^n \log \left(1 + D \frac{|x - x'|}{\beta^{n-j}} \right)$$

where D bounds $\frac{|T''|}{|T'|}$ on I. Then we have a constant $C > 1$ such that

$$\frac{\sup_{x \in TI_{i_n}} |\psi'_{i_1 \ldots i_n}(x)|}{\inf_{x \in TI_{i_n}} |\psi'_{i_1 \ldots i_n}(x)|} \leq C, \quad \forall i_1, \ldots, i_n, n > 0.$$

The property allows us to find a constant $K < +\infty$ such that

$$\sum_{i_1, \ldots, i_n} |\psi'_{i_1 \ldots i_n}(x)| \chi_{TI_{i_n}}(x) \leq K, \quad \forall x.$$

We conclude that there exists $D' > 0$ such that $|\mathcal{L}^n 1(x) - \mathcal{L}^n 1(x')| \leq$ $K \exp \left(|x - y| \frac{D'}{1 - \frac{1}{\beta}} \right)$ (where none of the bounds on the right hand side depends on n). We conclude that $\forall n \geq 1$

(1) the functions $\mathcal{L}^n 1$ are bounded in the supremum norm,
(2) the functions $\mathcal{L}^n 1$ are an equicontinuous family.

We construct a new family of averages

$$F_n(x) = \frac{1}{n} \sum_{k=0}^{n-1} \mathcal{L}^k 1(x), \qquad n \geq 0.$$

We again see that

(1) the functions F_n are bounded in the supremum norm,
(2) the functions F_n are an equicontinuous family.

By the Ascoli theorem, there must be a limit point $F_{n_r} \to f$ (≥ 0) in the continuous functions on each component of $[0, 1] - \{x_0, \ldots, x_k\}$ and since $\int \mathcal{L}^n 1 dx = 1$ we have $\int f d\lambda = \lim_{n \to \infty} \int F_{n_r} d\lambda = 1$. Moreover, we see that

$$\begin{aligned}
\mathcal{L}F_{n_r}(x) &= \sum_{y \in T^{-1}x} \frac{F_{n_r}(y)}{|T'(y)|} = \sum_{y \in T^{-1}x} \frac{1}{n_r} \sum_{k=0}^{n_r-1} \frac{\mathcal{L}^k 1(y)}{|T'(y)|} \\
&= \frac{1}{n_r} \sum_{k=0}^{n_r-1} \sum_{y \in T^{-1}x} \frac{\mathcal{L}^k 1(y)}{|T'(y)|} \\
&= \frac{1}{n_r} \sum_{k=0}^{n_r-1} \mathcal{L}^{k+1} 1(x) \\
&= F_{n_r}(x) - \frac{1}{n_r} \left(1(x) - \mathcal{L}^{n_r} 1(x) \right).
\end{aligned}$$

Letting $r \to +\infty$ we see that

$$\mathcal{L}f(x) = \sum_{y \in T^{-1}x} \frac{f(y)}{|T'(y)|} = f(x),$$

completing the proof.   ∎

DEFINITION. We say that T is aperiodic if there exists a positive number m such that $\lambda(T^{-m} I_i \cap I_j) > 0$, $\forall i, j > 0$.

THEOREM 12.5. *The absolutely continuous invariant measure μ in Proposition 12.4 is exact if $T : I \to I$ is aperiodic.*

PROOF. By Proposition 12.1 it suffices to show that for any set $A \in \mathcal{B}$ with $\mu(A) > 0$ and for which $T^n A \in \mathcal{B}$ for all $n \geq 0$ we have that $\lim_{n \to +\infty} \mu(T^n A) = 1$.

Given $\underline{i} = (i_1, \dots, i_n)$ we write $I_{\underline{i}} = \cap_{j=1}^{n-1} T^{-j+1} I_{i_j}$ if $\mathrm{int}\left(\cap_{j=1}^{n-1} T^{-j+1} I_{i_j}\right) \neq \emptyset$. As T is piecewise invertible on each atom I_i, we know that $T^n|_{I_{\underline{i}}}$ is a C^1-diffeomorphism. For all $I_{\underline{i}}$ and for all $n > 0$ we write $(T^n|_{I_{\underline{i}}})^{-1} = \psi_{\underline{i}}$. Let $x, y \in T^n I_{\underline{i}}$ $(= T I_{i_n})$; then it follows from the mean value theorem that

$$|\psi_{\underline{i}}(x) - \psi_{\underline{i}}(y)| = |\psi'_{\underline{i}}(\theta)||x - y|$$

for some $\theta \in I_{\underline{i}}$. From the above equality and the condition (i), the diameter $\mathrm{diam}\left(I_{\underline{i}}\right)$ of $I_{\underline{i}}$ decays exponentially fast (i.e., $\mathrm{diam}(I_{\underline{i}}) \leq \frac{1}{\beta^n}$). This implies that the partition $\mathcal{I} = \{I_i\}$ is a " generating partition". In particular, for any $\epsilon > 0$ we can choose a finite disjoint set of cylinders $\{I_{\underline{j}} : \underline{j} = (j_1, \dots, j_l)\}$, say, with $A \subset \cup_{\underline{j}} I_{\underline{j}}$ and $\mu\left(\left(\cup_{\underline{j}} I_{\underline{j}}\right) - A\right) < \epsilon$.

The following estimates will be useful in the rest of the proof.

(a) Given $\delta > 0$ there exists at least one cylinder $I_{\underline{j}}$ (where $\underline{j} = (j_1 \dots j_l)$, say) for which

$$\lambda(A \cap I_{\underline{j}}) \geq (1 - \delta)\,\lambda(I_{\underline{j}}). \tag{12.1}$$

Assume for a contradiction that this is not the case, then for all cylinders $I_{\underline{j}}$ we would have $\lambda(A \cap I_{\underline{j}}) \leq (1 - \delta)\lambda(I_{\underline{j}})$. We can extend this inequality to disjoint unions of cylinders, and then by approximation to arbitrary sets $B \in \mathcal{B}$ to get $\mu(A \cap B) \leq (1 - \delta)\mu(B)$. However, if we take $B = A$, then we get $\mu(A) \leq (1 - \delta)\mu(A)$, which contradicts $\mu(A) > 0$.

(b) We observe that there is a constant $C \geq 1$ such that for any cylinder $I_{\underline{i}}$

$$\sup_{x,y \in I_{\underline{i}}} \frac{|\psi'_{\underline{i}}(x)|}{|\psi'_{\underline{i}}(y)|} \leq C. \tag{12.2}$$

(This is usually referred to as *Renyi's condition*.)

From the change of variables formula we see that

$$
\begin{aligned}
\lambda(T^l I_{\underline{j}} \cap (T^l A)^c) &\le \int_{I_{\underline{j}} \cap A^c} |(T^l)'(x)| d\lambda(x) \\
&\le \left(\sup_{y \in I_{\underline{j}}} |(T^l)'(y)| \right) \lambda(I_{\underline{j}} \cap A^c) \\
&\le C \left(\inf_{y \in I_{\underline{j}}} |(T^l)'(y)| \right) \lambda(I_j \cap A^c) \quad \text{(using (12.2))} \\
&\le C \frac{\int_{I_j} |(T^l)'(x)| d\lambda(x)}{\lambda(I_j)} \lambda(I_j \cap A^c) \\
&\le C\delta\lambda(T^l I_j) \quad\quad\quad\quad\quad \text{(using (12.1))}.
\end{aligned}
$$

If $T^l(I_{\underline{j}}) = I$, then we could proceed directly to the end of the proof. However, since this need not be the case, we require the following sublemma.

SUBLEMMA 12.5.1. *There exist $S > 0$ and a finite disjoint union of cylinders I' of length S such that $I' \subset T^l(I_{\underline{j}})$ and $T^S(I') = I$.*

PROOF. Let $\{U_1, \dots, U_N\} = \{TI_1, \dots, TI_k\}$, where $N \le k$, denote the collection of images under T of the original intervals. The aperiodicity assumption implies that for each $1 \le j \le N$ there exists $0 < s_j < +\infty$ such that each U_i, $i = 1, \dots, N$, contains a cylinder $I^{(i,j)}_{m_1, \dots, m_{s_j}}$ satisfying $T^{s_j} I^{(i,j)}_{m_1, \dots, m_{s_j}} = U_j$. In particular, we see that $T^{s_i} U_i \supset T^{s_i} I^{(i,j)}_{m_1, \dots, m_{s_j}} = U_i$.

Let $T^l(I_{\underline{j}}) = U_i$. Setting $S = \prod_{j=1}^N s_j$ and $I' = \cup_{j=1}^N I^{(i,j)}_{m_1, \dots, m_{s_j}}$ allows us to have that $I' \subset T^l(I_{\underline{j}})$ and $T^S I' \supset \cup_{j=1}^N U_j = X$. ∎

We need only modify the previous argument to write

$$
\lambda(T^S(I' \cap (T^l A)^c)) \le D\delta
$$

for some uniform constant $D > 0$. Since $\lambda(T^S(I' \cap (T^l A)^c)) \ge 1 - \lambda(T^S(I' \cap T^l A))$, we see that

$$
\lambda(T^{l+S} A) > \lambda(T^S(I' \cap T^l A)) > 1 - D\delta.
$$

Since μ is absolutely continuous with respect to λ we conclude that $\mu(T^n A) \to 1$ as $n \to +\infty$. ∎

COROLLARY 12.5.1. *If $T : I \to I$ is aperiodic, then it is strong-mixing with respect to any absolutely continuous invariant measure. In particular, there exists a unique absolutely continuous invariant probability measure.*

PROOF. By Proposition 12.1 the exact measure μ is also strong mixing. By Proposition 11.2 it is also ergodic, and since no two distinct ergodic measures can be equivalent to Lebesgue measure (and thus each other) uniqueness follows.

PROPOSITION 12.6. μ *is equivalent to* λ.

PROOF. First we show the following fact:

$$\forall \epsilon > 0, \exists N(\epsilon) > 0 \text{ such that for each } x \in I, T^{-N(\epsilon)}x \text{ is } \epsilon\text{-dense in } I. \quad (12.3)$$

As we have already observed in Theorem 12.5, for $\forall I_{j_1 \ldots j_l}$ there exist a set of cylinders $\{I_{tm_1 \ldots m_{s_i}}^{(i)} : i = 1, \ldots N\}$ and $S > 0$ satisfying $T^S(\cup_{i=1}^{N} I_{tm_1 \ldots m_{s_i}}^{(i)}) = I$. Let $x \in I_{h_1 \ldots h_t}$. Then $\exists i$ s.t. $tm_1 \ldots m_{s_i} h_1 \ldots h_t$ is an admissible sequence and so $\psi_{tm_1 \ldots m_{s_i} h_1 \ldots h_t}(x) \in I_{tm_1 \ldots m_{s_i}} \subset T^l I_{j_1 \ldots j_l}$. Hence we have that $T^{-(l+s_i+t)}x \cap I_{j_1 \ldots j_l} \neq \emptyset$. Let $l = l(\epsilon)$ be a positive integer such that $\sup_{I_{j_1 \ldots j_l}} \operatorname{diam} I_{j_1 \ldots j_l} < \epsilon$. Then, each $I_{j_1 \ldots j_l}$ contains at least a point belonging to $T^{-(l+S)}x$. (Here we take $t = S - s_i$.) Choosing $N(\epsilon) = l(\epsilon) - S$, we have the fact (12.3).

It remains to show that f is bounded away from zero. Assume for a contradiction that $f(x) = 0$. Then since for all $n \geq 1, \mathcal{L}^n f(x) = \sum_{y \in T^{-n}x} \frac{f(y)}{|T^{n'}(y)|} = 0$, we see that $f(y) = 0$ whenever $T^n y = x$. By the property (12.3) the set of such points is dense. The continuity of f implies that f is identically zero, contradicting $\int f d\lambda = 1$. ∎

PROPOSITION 12.7. *For irreducible piecewise expanding Markov maps $T : I \to I$ the following condition is equivalent to strong mixing:*

$$\lambda \circ T^{-n}(A) \to \mu(A), \quad as \ n \to +\infty \quad (\forall A \in \mathcal{B}),$$

where λ is Lebesgue measure.

PROOF. It is enough to observe that

$$\lambda(T^{-n}A) = \int_I \chi_{T^{-n}A}(x) d\lambda(x) = \int \chi_A(T^n x) f(x)^{-1} d\mu(x)$$

$$\to (\int \chi_A(x) d(\mu(x)) \cdot (\int d\lambda(x)) = \mu(A).$$

∎

REMARK. Under the generating condition we can extend these results to multi-dimensional piecewise expanding Markov maps with countable infinite partitions.

Since the invariant density f is strictly positive, we can make the following definition.

DEFINITION. We define an operator $\hat{\mathcal{L}} : L^1(I) \to L^1(I)$ by $\hat{\mathcal{L}}(h) = \frac{1}{f}\mathcal{L}(fh)$ where $h \in L^1(X)$.

PROPOSITION 12.8. $\hat{\mathcal{L}}^*(\mu) = \mu$, i.e. the dual operator $\hat{\mathcal{L}}^*$ acting on measures (defined by $(\mathcal{L}^*\mu)(A) = \int \mathcal{L}\chi_A d\mu$) fixes μ.

PROOF. It is an immediate consequence of Sublemma 14.2.3 and the definition. ∎

THEOREM 12.9 (CONVERGENCE TO INVARIANT DENSITY). $\mathcal{L}^n(h) \to f\left(\int h d\lambda\right)$ uniformly for $\forall h \in C^0(I)$.

PROOF. Define $g = \frac{f(x)}{f(Tx)|T'(x)|}$. From Renyi's condition we have that there exists a uniform constant $D \geq 1$ such that $\forall x, x' \in U_k$

$$D(x, x') = \sup_{n \geq 1} \sup_{y \in T^{-n}x, y' \in T^{-n}x'} \prod_{i=1}^{n-1} \frac{|g(T^i y)|}{|g(T^i y')|}$$

is bounded above by D and furthermore

$$D(x, x') \to 1 \text{ as } |x - x'| \to 0.$$

An easy calculation shows that $\{\hat{\mathcal{L}}^n h : n \geq 0\}$ is equicontinuous on each component of $I - \partial \mathcal{V}$ for $\forall h \in C_0(I - \partial \mathcal{V})$. It follows from the definition of $\hat{\mathcal{L}}^n$ that $||\hat{\mathcal{L}}^n h||_\infty$ is bounded by $||h||_\infty$ and so the closure of $\{\hat{\mathcal{L}}^n h : n \geq 0\}$ in $C(I - \partial \mathcal{V})$ is compact. Hence there are a subsequence $\{n_i\} \to \infty$ $(i \to \infty)$ and $h^* \in C^0(I - \partial \mathcal{V})$ such that $\hat{\mathcal{L}}^{n_i} h \to h^*$ uniformly.

We can now show that any limit point of the sequence is a constant which, in particular, shows that the limit exists. Notice that $\min_{x \in I}(\hat{\mathcal{L}}^k h^*(x)) = \min_{x \in I}(h^*(x))$ for all $k \geq 0$. For any $k \geq 0$ choose $z \in I$ such that $\hat{\mathcal{L}}^k h^*(z) = \min_{x \in I} h^*(x)$. Then for all $y \in T^{-k}z$ we have that $h^*(y) = \min_{x \in I} h^*(x)$. In fact,

$$\hat{\mathcal{L}}^k h^*(z) = \sum_{T^k y = z} \left(g(y) \ldots g(T^{k-1}y)\right) h^*(y) \geq \min_{x \in I} h^*(x)$$

with equality if and only if $h^*(y) = \min_{x \in I} h^*(x)$, $\forall y \in T^{-n}x$. By (12.3) we see that the set of y such that $\exists k \geq 1$ with $T^k y = x$ is dense. Thus h^* is a constant function with value $\min_{x \in I} h^*(x)$ on a dense set, and thus by piecewise continuity is constant almost everywhere.

Moreover, this constant takes the value $\lim_{n \to +\infty} \int \hat{\mathcal{L}}^n h d\mu = \int h d\mu$. Replacing h by $\frac{h}{f}$ for $h \in C^0(I)$ and appealing to the definition of $\hat{\mathcal{L}}$ we get that

$$\mathcal{L}^n(h) = \mathcal{L}^n(f \cdot \frac{h}{f}) = f \cdot \hat{\mathcal{L}}^n(\frac{h}{f}) \to f\left(\int \frac{h}{f} \cdot f d\lambda\right) = f \cdot \left(\int h d\lambda\right)$$

uniformly as $n \to +\infty$.

■

12.3 Rohlin's entropy formula

In this section we want to give a formula for the entropy of an irreducible Markov piecewise expanding interval map $T : I \to I$ with respect to the unique absolutely continuous probability measure μ.

THEOREM 12.10 (ROHLIN ENTROPY FORMULA).

$$h_\mu(T) = \int \log |T'(x)| d\mu(x).$$

PROOF. The proof follows immediately from the string of statements (i)-(iv) below.

(i) By the chain rule we can write $\log |(T^N)'(x)| = \sum_{i=0}^{N-1} \log |T'(x)|$ for each $x \in I$, $N \geq 1$. Since the measure μ is ergodic (even exact) we can apply the Birkhoff ergodic theorem to deduce that

$$\frac{1}{N} \log |(T^N)'(x)| \to \int \log |T'(x)| d\mu(x) \text{ as } N \to +\infty.$$

(ii) Let $x \in I_{i_1,\ldots,i_N} = \cap_{j=1}^N T^{-(j-1)} I_{i_j}$; then using Renyi's condition we can estimate

$$\lambda(I_{i_1\ldots i_N}) = \int_{T^N I_{i_1\ldots i_N}} \frac{1}{|(T^N)'(\psi_{i_1\ldots i_N} z)|} d\lambda(z)$$

$$\leq C \left(\inf_{x \in I_{i_1\ldots i_N}} \frac{1}{|(T^N)'(x)|} \right) \lambda(T I_{i_N})$$

$$\leq C \left(\frac{1}{(T^N)'(x)} \right)$$

and

$$\lambda(I_{i_1\ldots i_N}) \geq \frac{1}{C} \left(\sup_{z \in T^n I_{i_1\ldots i_N}} \frac{1}{|(T^N)'(z)|} \right) \lambda(T(I_{i_N}))$$

$$\geq \frac{1}{C} \left(\min_{1 \leq i \leq k} \lambda(I_i) \right) \frac{1}{|(T^N)'(x)|}.$$

Thus we see that for any $x \in I$

$$-\lim_{N \to +\infty} \frac{1}{N} \log \lambda(I_{i_1\ldots i_N}) = \lim_{N \to +\infty} \frac{1}{N} \log |(T^N)'(x)|$$

(where $x \in I_{i_1 \ldots i_N}$).

(iii) Since the density f of the invariant measure is bounded from below and away from infinity, we see that

$$- \lim_{N \to +\infty} \frac{1}{N} \log \lambda(I_{i_1 \ldots i_N}) = - \lim_{N \to +\infty} \frac{1}{N} \log \mu(I_{i_1 \ldots i_N}).$$

(iv) Finally, we claim that

$$- \lim_{N \to +\infty} \frac{1}{N} \log \mu(I_{i_1 \ldots i_N}) = h_\mu(T).$$

This is an application of the Shannon-McMillan-Brieman theorem to interval maps, whose proof we present in the next section.   ∎

12.4 The Shannon-McMillan-Brieman theorem

We now give an application of entropy to describe the asymptotic size of elements in partitions.

Let $\alpha = \{A_1, A_2, \ldots\}$ be a measurable partition of the space $(X, \mathcal{B})$, i.e. $X = \cup_{i=1}^n A_i$ and $A_i \cap A_j = \emptyset$ for $i \neq j$ (up to a set of zero μ-measure).

For each $n \geq 1$ we consider the new partition $\alpha_n = \vee_{i=0}^\infty T^{-i}\alpha$. For almost all $x \in X$ we can choose a unique element $A_n(x) \in \alpha_n$ with $x \in A_n(x)$.

THEOREM 12.11 (SHANNON-MCMILLAN-BRIEMAN THEOREM). *Let $T : X \to X$ be a measure preserving transformation of a probability space $(X, \mathcal{B}, \mu)$. Let α be a partition. For almost all $x \in X$ we have that*

$$- \frac{\log \mu(A_n(x))}{n} \to E(f|\mathcal{I})(x)$$

as $n \to +\infty$, where $f(x) = I(\alpha| \vee_{n=1}^\infty T^{-i}\alpha)(x)$ and $\mathcal{I}$ is the sigma-algebra generated by the T-invariant sets $T^{-1}B = B$.

COROLLARY 12.11.1. *If T is ergodic then for almost all $x \in X$*

$$- \frac{\log \mu(A_n(x))}{n} \to h(T, \alpha) \text{ as } n \to +\infty.$$

If α is a generating partition then

$$- \frac{\log \mu(A_n(x))}{n} \to h_\mu(T) \text{ as } n \to +\infty.$$

PROOF. Assuming the theorem, the ergodicity of the measure and the T-invariance of the limit imply that it is a constant. Integrating therefore gives that the limit is

$$\int E(f|\mathcal{I})d\mu = \int f d\mu = H(\alpha|\vee_{n=1}^{\infty}T^{-n}\alpha) = h(T,\alpha).$$

PROOF OF THEOREM 12.11. We first observe that

$$I(\vee_{i=0}^{n-1}T^{-i}\alpha)(x) = -\log\mu\,(A_n(x))$$

Using the basic identities for the information function we see that

$$\begin{aligned}
&I(\vee_{i=0}^{n-1}T^{-i}\alpha)\\
&= I(\alpha|\vee_{i=1}^{n-1}T^{-i}\alpha) + I(\vee_{i=1}^{n-1}T^{-i}\alpha)\\
&= I(\alpha|\vee_{i=1}^{n-1}T^{-i}\alpha) + I(\alpha|\vee_{i=1}^{n-2}T^{-i}\alpha)T\\
&\quad + \ldots + I(\alpha|T^{-1}\alpha)T^{n-2} + I(\alpha)T^{n-1}.
\end{aligned} \tag{12.4}$$

We see from (12.4) that (almost everywhere)

$$\begin{aligned}
&\limsup_{n\to+\infty}\frac{1}{n}|I(\vee_{i=0}^{n-1}T^{-i}\alpha) - E(f|\mathcal{I})|\\
&\leq \limsup_{n\to+\infty}\frac{1}{n}|I(\vee_{i=0}^{n-1}T^{-i}\alpha) - \sum_{i=0}^{n-1}fT^i|\\
&\quad + \limsup_{n\to+\infty}|\frac{1}{n}\sum_{i=0}^{n-1}fT^i - E(f|\mathcal{I})|
\end{aligned} \tag{12.5}$$

(using the triangle inequality). By the Birkhoff ergodic theorem (Theorem 10.6) we know that

$$\lim_{n\to+\infty}\frac{1}{n}|\sum_{i=0}^{n-1}fT^i - E(f|\mathcal{I})| = 0$$

(almost everywhere) and thus the second term on the right hand side of (12.5) vanishes.

We can next write from (12.5) that

$$\begin{aligned}
&\frac{1}{n}\left|I(\vee_{i=0}^{n-1}T^{-i}\alpha) - \sum_{i=0}^{n-1}fT^i\right|\\
&\leq \frac{1}{n}\sum_{i=0}^{n-1}|I(\alpha|\vee_{j=1}^{n-i}T^{-j}\alpha)T^i - I(\alpha|\vee_{j=1}^{\infty}T^{-j}\alpha)T^i|
\end{aligned}$$

(using also the definition of f). For $N \geq 1$ we define

$$F_N(x) := \sup_{N \leq i \leq n} |I(\alpha| \vee_{j=1}^{n-i} T^{-j}\alpha)(x) - I(\alpha| \vee_{j=1}^{\infty} T^{-j}\alpha)(x)|$$

and then upon fixing $N \geq 1$ we see that

$$\frac{1}{n}\left|I(\vee_{i=0}^{n-1} T^{-i}\alpha) - \sum_{i=0}^{n-1} fT^i\right|$$

$$\leq \left(\frac{F_N T^n + F_N T^{n-1} + \ldots + F_N T^{n-N}}{n}\right) \qquad (12.6)$$

$$+ \left(\frac{\sum_{i=0}^{N-1} |I(\alpha| \vee_{j=1}^{n-i} T^{-j}\alpha)T^i - I(\alpha| \vee_{j=1}^{\infty} T^{-j}\alpha)T^i|}{n}\right)$$

We can bound the second term on the right hand side of (12.6) by

$$\frac{1}{n}\left|\sum_{i=0}^{N-1} |I(\alpha| \vee_{j=1}^{n-i} T^{-j}\alpha)T^i - I(\alpha| \vee_{j=1}^{\infty} T^{-j}\alpha)T^i|\right|$$

$$\leq \frac{N}{n}\left(\sup_{k \geq N} \left(I(\alpha| \vee_{j=1}^{\infty} T^{-j}\alpha) + I(\alpha| \vee_{j=1}^{k} T^{-j}\alpha)\right)\right)$$

which tends to 0 (almost everywhere) as $n \to +\infty$.

We now turn to the first term on the right hand side of (12.6). We observe that by the Birkhoff ergodic theorem

$$\limsup_{n \to +\infty} \left(\frac{F_N T^n + F_N T^{n-1} + \ldots + F_N T^{n-N}}{n}\right) = E(F_N|\mathcal{I}).$$

Notice that $F_N \geq F_{N+1}$ and so

$$E(F_N|\mathcal{I}) \geq E(F_{N+1}|\mathcal{I}) \geq 0$$

(since $E(.|\mathcal{I})$ is a positive operator). Since $E(F_N|\mathcal{I}) \to 0$ (and is dominated by an integrable function) then

$$\lim_{N \to +\infty} \int E(F_N|\mathcal{I})d\mu = \lim_{N \to +\infty} \int F_N d\mu = 0.$$

This completes the proof.   ∎

12.5 Comments and references

A good reference for more information on exactness is Rohlin's original paper [2].

Without the Markov assumption (but still assuming the uniform expansion property) the existence of an absolutely continuous invariant measure follows from the work of Lasota and Yorke [1].

There is an alternative proof of the Shannon-McMillan-Brieman theorem given in [3, 5.2]

References

1. A. Lasota and J. Yorke, *On the existence of invariant measures for piecewise mono-tonic transformations*, Trans. Amer. Math. Soc. **86** (1973), 481-488.
2. V. Rohlin, *Exact endomorphisms of lebesgue space*, Amer. Math. Soc. Transl. (2) **39** (1964), 1-36.
3. D. Rudolph, *Ergodic Theory on Lebesgue spaces*, O.U.P., Oxford, 1994.

CHAPTER 13

FIXED POINTS FOR
HOMEOMORPHISMS OF THE ANNULUS

13.1 Fixed points for the annulus

Let $A = \mathbb{R}/\mathbb{Z} \times [0,1]$ be a closed annulus. Assume that $T : A \to A$ is a homeomorphism that preserves the two boundary circles (i.e. $T(\mathbb{R}/\mathbb{Z} \times \{0\}) = \mathbb{R}/\mathbb{Z} \times \{0\}$ and $T(\mathbb{R}/\mathbb{Z} \times \{1\}) = \mathbb{R}/\mathbb{Z} \times \{1\}$).

DEFINITION. We say that $T : A \to A$ is *area preserving* if the Haar-Lebesgue measure λ is T-invariant.

REMARK. We say that $T : A \to A$ is *conservative* if there are no wandering sets of positive measure. This condition will suffice for most of the results of this section.

EXAMPLE. For any pair of values $0 < \alpha, \beta < 1$ consider the map $T : A \to A$ given by $T(x,y) = (x + \alpha y + \beta(1-y), y)$. To see that this is area preserving we can write this affine transformation as $T(x,y) = (\beta, 0) + B(x,y)$ where $B = \begin{pmatrix} 1 & \alpha - \beta \\ 0 & 1 \end{pmatrix}$. Since $\det(B) = 1$ we see that T is area preserving.

DEFINITION. An ϵ-chain (for $T : A \to A$) from (x,y) to (w,z) is a sequence of points

$$(x,y) = (x_0, y_0), (x_1, y_1), \ldots, (x_n, y_n) = (w, z) \in A$$

such that

$$d(T(x_i, y_i), (x_{i+1}, y_{i+1})) < \epsilon \text{ for } i = 0, \ldots, n-1.$$

We can use the same notation for ϵ-chains for the lift $\hat{T} : \mathbb{R} \times [0,1] \to \mathbb{R} \times [0,1]$.

These are finite versions of the pseudo-orbits introduced in chapter 5.

LEMMA 13.1. *If* $(x,y) = (x_0, y_0), (x_1, y_1), \ldots, (x_n, y_n) = (w, z)$ *is an ϵ-chain from* (x,y) *to* (w,z) *and* $(w,z) = (w_0, z_0), (w_1, z_1), \ldots, (w_n, z_n) = (u,v)$ *is an ϵ-chain from* (w,z) *to* (u,v) *then defining* $(x_{n+i}, y_{n+i}) = (w_i, z_i)$ *for* $0 \le i \le m$ *makes* $(x_0, y_0), \ldots, (x_n, y_n), (x_{n+1}, y_{n+1}), \ldots, (x_{n+m}, y_{n+m})$ *an ϵ-chain from* (x,y) *to* (u,v).

PROOF. This is immediate from the definitions.   ∎

DEFINITION. We say that a point (x, y) is *chain recurrent* if for each $\epsilon > 0$ we can find an ϵ-pseudo-orbit from (x, y) to itself.

We say that $T : A \to A$ is *chain recurrent* if for every $(x, y), (w, z) \in A$ and every $\epsilon > 0$ we can find a finite ϵ-pseudo-orbit from (x, y) to (w, z).

LEMMA 13.2. *Let $T : A \to A$ be an area preserving homeomorphism; then*

   (i) *every point $(x, y) \in A$ is chain recurrent,*

   (ii) *$T : A \to A$ is chain recurrent.*

PROOF. (i) Fix $\epsilon > 0$ and then by (uniform) continuity we can choose $\frac{\epsilon}{2} > \delta > 0$ such that whenever $|(x, y) - (u, v)| < \delta$ then $|T(x, y) - T(u, v)| < \frac{\epsilon}{2}$. Let us choose a finite cover of δ-balls

$$A \subset \cup_{i=1}^{N} B((z_i, w_i), \delta) \text{ where } (z_1, w_1), \dots, (z_N, w_N) \in A.$$

Since T is area preserving we have for each $i = 1, \dots, N$ that we can choose n_i such that $T^{-n_i} B((z_i, w_i), \delta) \cap B((z_i, w_i)\delta) \neq \emptyset$. If we choose (u_i, v_i) in this intersection then

$$\begin{cases} (x_0, y_0) = (u_i, v_i), \\ (x_1, y_1) = T(u_i, v_i), \\ \vdots \\ (x_j, y_j) = T^j(u_i, v_i), \\ \vdots \\ (x_{n_i-1}, y_{n_i-1}) = T^{(n_i-1)}(u_i, v_i), \\ (x_{n_i}, y_{n_i}) = (u_i, v_i) \end{cases}$$

gives an $\frac{\epsilon}{2}$-chain from (z_i, w_i) to itself since we observe that

   (a) $|T((x_0, y_0)) - T(z_i, w_i)| < \frac{\epsilon}{2}$ (since $|(x_0, y_0) - (z_i, w_i)| = |(u_i, v_i) - (z_i, w_i)| < \delta$),

   (b) $T(x_j, y_j) = T(T^j(u_i, v_j)) = T^{j+1}(u_j, v_j)$ for $j = 1, \dots, n_i - 2$; and

   (c) $T(x_{n_i-1}, y_{n_i-1}) = T^{n_i}(u_j, v_j) \in B((z_i, w_i), \delta) \subset B((z_i, w_i), \frac{\epsilon}{2}))$.

For any $(z, w) \in A$ we can choose some (z_i, w_i) $(i = 1, \dots, N)$ which is δ-close to (z, w) (i.e. $|(z_i, w_i) - (z, w)| < \delta$). We then see that the above $\frac{\epsilon}{2}$-chain from (z_i, w_i) to itself also serves as an ϵ-chain from (z, w) to itself on replacing both (x_0, y_0) and (x_{n_i}, y_{n_i}) by (z, w). To see this observe that:

   (d) $|T(z, w) - (x_1, y_1)| \leq |T(z, w)) - T(z_i, w_i)| + |T(z_i, w_i) - T(x_1, y_1)| < \frac{\epsilon}{2} + \frac{\epsilon}{2} = \epsilon$ (since $|(z_i, w_i) - (z, w)| < \delta$ implies that $|T(z_i, w_i) - T(z, w)| < \frac{\epsilon}{2}$),

   (e) $|T(x_{n_i}, y_{n_i}) - (z, w)| \leq |T(x_{n_i}, y_{n_i}) - (z_i, w_i)| + |(z_i, w_i) - (z, w)| \leq \frac{\epsilon}{2} + \delta < \epsilon)$.

(ii) We may choose a sequence $(z_{i_0}, w_{i_0}), (z_{i_1}, w_{i_1}), \ldots, (z_{i_n}, w_{i_n})$ such that

$$\begin{cases} (x,y) \in B((z_{i_0}, w_{i_0}), \delta), \\ B((z_{i_j}, w_{i_j}), \delta) \cap B((z_{i_{j+1}}, w_{i_{j+1}}), \delta) \neq \emptyset \text{ for } 0 \leq j \leq n-1, \\ (u,v) \in B((z_{i_n}, w_{i_n}), \delta). \end{cases}$$

If we write in succession the ϵ-pseudo-orbit from (z_{i_0}, w_{i_0}) to (z_{i_0}, w_{i_0}), and then the $\frac{\epsilon}{2}$-pseudo-orbit from (z_{i_1}, w_{i_1}) to (z_{i_1}, w_{i_1}), etc. until we get to the $\frac{\epsilon}{2}$-pseudo-orbit from (z_{i_n}, w_{i_n}) to (z_{i_n}, w_{i_n}) then resulting concatenated sequence is an ϵ-chain from (x,y) to (u,v) (cf. Lemma 13.1). ∎

The following result tells us that there is no distinction between the existence of fixed points and periodic points for the map $\hat{T} : \mathbb{R} \times [0,1] \to \mathbb{R} \times [0,1]$.

BROUWER PLANE THEOREM. *If $\hat{T} : \mathbb{R} \times [0,1] \to \mathbb{R} \times [0,1]$ has a periodic point (i.e. $\exists n \neq 1, \hat{T}^n(x,y) = (x,y)$) then $\hat{T}$ has a fixed point (i.e. $\exists T(u,v) = (u,v)$).*

This is a classical result. We sketch the proof in the final section. (For detailed proofs we refer to [2], [1].)

THEOREM 13.3 (POINCARÉ-BIRKHOFF). *Assume that $T : A \to A$ is an area preserving homeomorphism and that the rotation numbers ρ_0 and ρ_1 for $T : \mathbb{R}/\mathbb{Z} \times \{0\} \to \mathbb{R}/\mathbb{Z} \times \{0\}$ and $T : \mathbb{R}/\mathbb{Z} \times \{0\} \to \mathbb{R}/\mathbb{Z} \times \{0\}$, respectively, satisfy either $\rho_0 < 0 < \rho_1$ or $\rho_1 < 0 < \rho_0$. Then there exists a fixed point for T.*

REMARK. This result is also known as *Poincaré's last geometric theorem*. Usually the statement involves the existence of two distinct fixed points.

Theorem 13.3 is a special case of the following more general result.

THEOREM 13.4 (FRANKS). *Assume that $T : A \to A$ is chain recurrent and that the rotation numbers ρ_0 and ρ_1 for $T : \mathbb{R}/\mathbb{Z} \times \{0\} \to \mathbb{R}/\mathbb{Z} \times \{0\}$ and $T : \mathbb{R}/\mathbb{Z} \times \{0\} \to \mathbb{R}/\mathbb{Z} \times \{0\}$, respectively, satisfy either $\rho_0 < 0 < \rho_1$ or $\rho_1 < 0 < \rho_0$. Then there exists a fixed point for T.*

PROOF OF THEOREM 13.3 (ASSUMING THEOREM 13.4). By Lemma 13.2 the hypothesis that T is area preserving implies that T is chain recurrent. The results follows immediately from Theorem 13.4. ∎

It remains to prove Theorem 13.4.

PROOF OF THEOREM 13.4. The proof of the theorem will be conveniently divided into the following sublemmas.

SUBLEMMA 13.4.1. *Let* $\hat{T} : \mathbb{R} \times [0,1] \to \mathbb{R} \times [0,1]$ *be the lift of a chain recurrent homeomorphism* $T : A \to A$. *There are four possibilities:*

(i) $\hat{T} : \mathbb{R} \times [0,1] \to \mathbb{R} \times [0,1]$ *has a chain recurrent point,*

(ii) *all points move to the right (i.e.* $\forall (x,y) \in \mathbb{R} \times [0,1]$ *if* $(x^{(n)}, y^{(n)}) := \hat{T}^n(x,y)$ *then* $\lim_{n \to +\infty} x^{(n)} = +\infty$*),*

(iii) *all points move to the left (i.e.* $\forall (x,y) \in \mathbb{R} \times [0,1]$ *we have* $\lim_{n \to +\infty} x^{(n)} = -\infty$*), or*

(iv) $\forall M > 0$, $\exists (x,y), (w,z) \in \mathbb{R} \times [0,1]$ $\exists n, m \geq 1$ *with* $(x^{(n)} - x) < -M$ *and* $(w^{(n)} - w) > M$.

The following result shows that case (iv) is actually redundant.

SUBLEMMA 13.4.2. *Case (iv) implies case (i).*

SUBLEMMA 13.4.3. *Let* $\hat{T} : \mathbb{R} \times [0,1] \to \mathbb{R} \times [0,1]$ *be a lift of* $T : A \to A$. *Assume that* $\exists (x,y) \in \mathbb{R} \times [0,1]$, $\forall \epsilon > 0$, $\exists (x_i, y_i)$, $i = 0, \ldots, N$, *such that* $(x_0, y_0) = (x_n, y_n) = (x,y)$ *and* $|\hat{T}(x_i, y_i) - (x_{i+1}, y_{i+1})| < \epsilon$ *for* $i = 0, \ldots, n - 1$. *Then there exists a fixed point for* $\hat{T} : \mathbb{R} \times [0,1] \to \mathbb{R} \times [0,1]$.

Assuming these sublemmas the proof of Theorem 13.4 is now a simple matter. By the area preserving hypothesis $T : A \to A$ is chain recurrent and Sublemma 13.4.1 applies.

By the hypotheses on the rotation numbers of the boundaries, points on the two boundaries move in opposite directions. Thus we see that cases (ii) and (iii) are eliminated. If (i) holds then T has a chain recurrent point. If (iv) holds than by Sublemma 13.4.2 this again leads to the same conclusion, that there exists a chain recurrent point.

Finally, by Sublemma 13.4.3 the existence of a chain recurrent point implies the existence of a fixed point. ∎

We are only left with the chore of proving the sublemmas.

PROOF OF SUBLEMMA 13.4.1. Let us assume that (iv) fails, then to prove the sublemma we need to show that either (i), (ii) or (iii) holds.

Let us assume that (iv) fails because $\exists M > 0$, $\forall (x,y) \in \mathbb{R} \times [0,1]$, $\forall n \geq 1$ we have that $x_m^1 - x^1 \geq -M$. We have two possibilities.

Firstly, if for some $(x,y) \in \mathbb{R} \times [0,1]$ we have that the sequence $x^{(n)} - x$, $n \geq 1$, is also bounded above then the sequence $(\hat{T}^i(x,y))_{i=0}^\infty$ is confined to a bounded region of $\mathbb{R} \times [0,1]$ and so must have an accumulation point (x^*, y^*), say. However, for any $\delta > 0$ we need only choose $n' \geq n \geq 1$ with $|\hat{T}^{n'}(x,y) - (x^*, y^*)| < \frac{\delta}{2}$ and $|\hat{T}^n(x,y) - (x^*, y^*)| < \frac{\delta}{2}$ and then the sequence $((x^{(i)}, y^{(i)}))_{i=n}^{n'}$ is a δ-chain from (x^*, y^*) back to (x^*, y^*). Thus (x^*, y^*) is a chain recurrent point and we are in case (i).

The second possibility is that $\forall (x,y) \in \mathbb{R} \times [0,1]$ the sequence $x^{(n)} - x \geq -M$ ($n \geq 0$) is unbounded (i.e. $\hat{T}^n$ may move points arbitrarily far to the right, but never to the left). In particular, for any $C > 0$ we can choose N with $x^{(N)} - x \geq C$. Thus if $n \geq N$ then $x^{(n)} - x = \left(x^{(N)}\right)^{(n-N)} - x^{(N)} + (x^{(N)} - x) \geq -M + C$. i.e. $\lim_{n \to +\infty} x^{(n)} = +\infty$ (thus we are in case (ii)).

If we had assumed that (iv) failed because the second condition in (iv) was not met then a similar argument would have given that we are in either case (i) or case (iii).

∎

PROOF OF SUBLEMMA 13.4.2. We need a preliminary observation. Consider any two points $(u, v), (s, t) \in A$; then by Lemma 13.2 (i) we can find an ϵ-chain $(u_i, v_i)_{i=0}^{n}$ from (u, v) to (s, t). Lifting this chain to $\mathbb{R} \times [0, 1]$ we get that there is an ϵ-chain in $\mathbb{R} \times [0, 1]$ from (u, v) to $(s + r, t)$, say, for some $r \in \mathbb{Z}$. In addition, n can be bounded above by a bound D, say, depending only on ϵ and not on the choice of (u, v) and (s, t).

Returning to the proof of Sublemma 13.4.2, assuming property (iv) let us take $M > 4D$, then let $(x, y), (w, z) \in \mathbb{R} \times [0, 1]$ be the two points described in its statement.

(a) Given any point $(u, v) \in \mathbb{R} \times [0, 1]$ we can construct an ϵ-chain from (u, v) to $(x + r_1, y)$, for some $r_1 \in \mathbb{Z}$, by the above observation (with $|r_1| \leq D$).

(b) We can construct an ϵ-chain in $\mathbb{R} \times [0, 1]$ from $(x + r_1, y)$ to $(x^{(n)} + r_2, y^{(n)})$, for some $r_2 \in \mathbb{Z}$, by taking the lift of the orbit sequence $\left(x^{(i)}, y^{(i)}\right)_{i=0}^{n}$. By hypotheses, $r_2 \geq 4D$.

(c) We can construct an ϵ-chain in $\mathbb{R} \times [0, 1]$ from $(x^{(n)} + r_2, y^{(n)})$ to $(u + r_2, v)$, for some $r \in \mathbb{Z}$, by the above observation (with $|r_2| \leq D$).

Thus by Lemma 13.1 we can concatenate these to get an ϵ-chain from (u, v) to $(u, v) + (r, 0)$ with $r > 2D$

A similar argument (using the second part of property (iv)) shows that there is an ϵ-chain from z to $z - (s, 0)$, say, for some $s \in \mathbb{Z}$.

If $s = r$ then we can use Lemma 13.1 to combine the ϵ-chain from z to $z + (r, 0)$ with the ϵ-chain from $z + (r, 0)$ to $z + (r - s, 0) = z$ to get an ϵ-chain from z to itself. If $r \neq s$, we can repeat s times the ϵ-chain (applying Lemma 13.1 repeatedly) from z to $z + (r, 0)$ (to get an ϵ-chain from z to $z + (rs, 0)$) followed by r times the ϵ-chain from z to $z - (s, 0)$ (applying Lemma 13.1 repeatedly) to get from $z + (rs, 0)$ to z.

∎

PROOF OF SUBLEMMA 13.4.3. We shall first show that for each $\delta > 0$ we can find a homeomorphism $S : \mathbb{R} \times [0, 1] \to \mathbb{R} \times [0, 1]$ such that S has a fixed point and $\sup_{(x,y) \in \mathbb{R} \times [0,1]} |\hat{T}(x, y) - S(x, y)| < \frac{\delta}{2}$.

By hypothesis, we can choose a $\frac{\delta}{4}$-pseudo-orbit (x_i, y_i), $i = 0, \ldots, N$, from (x, y) to itself. We introduce a homeomorphism $h : \mathbb{R} \times [0, 1] \to \mathbb{R} \times [0, 1]$ such that $h(\hat{T}(x_i, y_i)) = (x_{i+1}, y_{i+1})$ and $\sup_{(x,y) \in \mathbb{R} \times [0,1]} |h(x, y) - (x, y)| < \frac{\delta}{2}$.

(Intuitively, this seems easy, although in practice it is harder to write down details.)

If we define $g(x, y) = (h \circ \hat{T})(x, y)$ then we can arrange that

$$\sup_{(x,y) \in \mathbb{R} \times [0,1]} |\hat{T}(x, y) - S(x, y)| < \frac{\delta}{2} \text{ and } S^n(x, y) = (x, y).$$

Thus there exists a periodic point for S and therefore by Brouwer's theorem there is a fixed point for S.

We observe that if we assume for a contradiction that $\hat{T}$ did *not* have a fixed point then (by compactness of A) the same would be true for any sufficiently close homeomorphim S. This contradicts the above construction.

Finally, this fixed point for $\hat{T}$ projects to a fixed point for $T : A \to A$.

13.2 Outline proof of Brouwer's theorem

In the previous section we made use of a classical (but not standard) result of Brouwer. In this section we shall ouline the main ideas in the proof.

OUTLINE PROOF. The proof has two distinct parts:

(i) If $T : \mathbb{R}^2 \to \mathbb{R}^2$ is a homeomorphism with a periodic point of prime period $n \geq 3$ then there exists a homeomorphism $T' : \mathbb{R}^2 \to \mathbb{R}^2$ with either a fixed point or a periodic point of prime period at most 2 *and the two homeomorphisms have the same set of fixed points.*

(ii) If $T : \mathbb{R}^2 \to \mathbb{R}^2$ has a periodic point of prime period 2 then there exists a fixed point.

Part (i) is proved by an iterative method. Specifically, If $T : \mathbb{R}^2 \to \mathbb{R}^2$ is a homeomorphism with a periodic point $T^n x = x$ of prime period $n \geq 3$ then one shows there exists a homeomorphism $T' : \mathbb{R}^2 \to \mathbb{R}^2$ with a periodic point of prime period at most $n - 1$ and the two homeomorphisms have the same set of fixed points.

To see this, consider the family of balls $B(x, \epsilon)$ about x, and their images $T(B(x, \epsilon))$ as neighbourhoods of Tx (cf. Figure 13.1). We choose the *smallest* $\epsilon > 0$ such that $\text{cl}(B(x, \epsilon)) \cap T(\text{cl}(B(x, \epsilon))) \neq \emptyset$. We can choose a point $z \in \text{cl}(B(x, \epsilon)) \cap T(\text{cl}(B(x, \epsilon)))$ and a path γ in $B(x, \epsilon)$ from $T^{-1}z$ to z (passing through x). By construction, this path γ has the property that $\gamma \cap T(\gamma) = \emptyset$. Since $x \in \gamma$ we see that $T^n(\gamma) \cap \gamma \neq \emptyset$ and so the "first" intersection $y \in T^k(\gamma) \cap \gamma$ ($n \geq k \geq 2$) gives rise to a simple closed curve containing $T(\gamma) \cup T^2(\gamma) \cup \ldots \cup T^{n-1}(\gamma)$. The homeomorphism T can be changed in a continuous way, or *isotopied* (but only in a small neighbourhood of this closed curve), to $T' : \mathbb{R}^2 \to \mathbb{R}^2$ so that z becomes a point of period $k - 1$ for T'.

Furthermore, since T can have no fixed points on the simple closed curve then it has some neighbourhood U in which the same is true. If we arrange

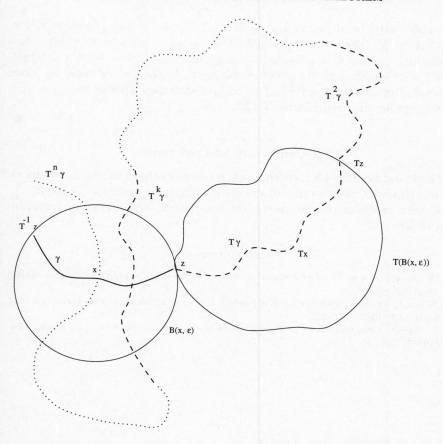

FIGURE 13.1. The proof of Brouwer's theorem

that T' differs from T only in this neighbourhood, then they have the same set of fixed points in $\mathbb{R}^2$.

Part (ii) is proved using some elementary topology. Assume that $T :$ $\mathbb{R}^2 \to \mathbb{R}^2$ has a periodic point $T^2 x = x$ (with x and Tx distinct). By "adding the fixed point at infinity" ∞ this corresponds to a homeomorphism on the standard 2-sphere S^2. By "blowing-up" the points x and Tx into circles we finally get a corresponding homeomorphism $\hat{T} : A \to A$ on a closed annulus $A = S^1 \times [0,1]$, say, which preserves orientation (since $T : \mathbb{R}^2 \to \mathbb{R}^2$ preserved orientation) and interchanges the two boundary components (since T interchanged x and Tx). Finally, the universal cover of the annulus is $X = \mathbb{R} \times [0,1]$, and the lift $\tilde{T} : X \to X$ interchanges the two sides $\mathbb{R} \times \{0\}$ and $\mathbb{R} \times \{1\}$. Let $\pi : X \to A$ be the covering projection and let $g : X \to X$ be a generator for the covering group (isomophic to $\mathbb{Z}$).

A simple application of the familiar Brouwer fixed point theorem gives that there exist fixed points $\tilde{T}z_0 = z_0$ for $\tilde{T} : X \to X$ and $(\tilde{T}g)z_1 = z_1$ for $(\tilde{T}g) : X \to X$. There are two possibilities: either $\pi(z_0) \neq \infty$ or $\pi(z_0) = \infty$. In the first case, $\pi(z)$ corresponds to a genuine fixed point for the original map $T : \mathbb{R}^2 \to \mathbb{R}^2$, and the proof is complete. In the second case, we have that $\pi(z_1)$ is distinct from $\pi(z_0) = \infty$ and so it corresponds to a genuine fixed point for the original map $T : \mathbb{R}^2 \to \mathbb{R}^2$.

∎

13.3 Comments and references

The Poincaré-Birkhoff theorem usual guarantees the existence of two fixed points. However, the theorem of Franks has much weaker hypotheses [3].

Modern proofs of the Brouwer plane translation theorem can be found in [1] and [2].

References

1. M. Brown, *A new proof of Brouwer's lemma on translation arcs*, Houston J. Math. **10** (1984), 35-41.
2. A. Fathi, *An orbit closing proof of Brouwer's lemma on translation arcs*, L'enseignment Math. **33** (1987), 315-322.
3. J. Franks, *Generalisations of the Poincaré-Birkhoff theorem*, Annals of Math. **128** (1988), 139-151.

CHAPTER 14

THE VARIATIONAL PRINCIPLE

We introduced in chapter 3 the topological entropy $h(T)$ of a continuous map $T : X \to X$ of a metric space X and in chapter 8 the entropy $h_\mu(T)$ of a T-invariant probability measure μ. In this chapter we show that these two notions are closely related.

14.1 The variational principle for entropy

The main result of this chapter is the following.

THEOREM 14.1 (VARIATIONAL PRINCIPLE). *Let $T : X \to X$ be a continuous map on a compact metric space.*

(1) *For any T-invariant probability measure μ we have that $h_\mu(T) \leq h(T)$.*

(2) *$h(T) = \sup\{h_\mu(T)\colon \mu$ is a T-invariant probability measure$\}$.*

14.2 The proof of the variational principle

The proof we give is due to Misiurewicz [1]. Recall that the topological entropy of a cover $\mathcal{U}$ is $H(\mathcal{U}) = \log N(\mathcal{U})$ and the entropy of a partition α with respect to μ is $H_\mu(\alpha) = -\sum_{A \in \alpha} \mu(A) \log \mu(A)$.

PROOF OF (1). Fix a finite measurable partition $\alpha = \{A_1, \ldots, A_k\}$ for X. Given $\epsilon > 0$, say, we want to "improve" this partition by choosing a family of closed sets $\hat{A}_1, \ldots, \hat{A}_k$ such that

(1) $\hat{A}_i \subset A_i$, $i = 1, \ldots, k$, and

(2) $\mu(A_i - \hat{A}_i) < \epsilon$,

and then defining a new partition $\hat{\alpha} = \{\hat{A}_1, \ldots, \hat{A}_k, V\}$, where $V = X - \left(\cup_{i=1}^k \hat{A}_i\right)$.

We can consider an open cover for X defined by

$$\mathcal{U} = \left\{ \hat{A}_1 \cup V, \ldots, \hat{A}_k \cup V \right\}$$

If we compare the open covers $\vee_{i=0}^{n-1} T^{-i} \mathcal{U}$ and the partitions $\vee_{i=0}^{n-1} T^{-i} \hat{\alpha}$ then we see that

$$N\left(\vee_{i=0}^{n-1} T^{-i} \hat{\alpha}\right) \leq 2^n N\left(\vee_{i=0}^{n-1} T^{-i} \mathcal{U}\right), \quad n \geq 1 \qquad (14.1)$$

(where we recall that $N(\vee_{i=0}^{n-1}T^{-i}\mathcal{U})$ is the number of elements in a minimal subcover for $\vee_{i=0}^{n-1}T^{-i}\mathcal{U}$ and $N(\vee_{i=0}^{n-1}T^{-i}\hat{\alpha})$ is the number of non-trivial elements in $\vee_{i=0}^{n-1}T^{-i}\hat{\alpha}$).

SUB-LEMMA 14.1.1. $H_\mu(\vee_{i=0}^{n-1}T^{-i}\hat{\alpha}) \leq \log N\left(\vee_{i=0}^{n-1}T^{-i}\hat{\alpha}\right).$

PROOF. Assume that $\vee_{i=0}^{n-1}T^{-i}\hat{\alpha} = \{C_1,\dots,C_N\}$; then we can write $H_\mu(\vee_{i=0}^{n-1}T^{-i}\hat{\alpha}) = -\sum_{i=1}^{N}\mu(C_i)\log\mu(C_i).$ ∎

We can use Sub-lemma 14.1 to bound

$H_\mu(\vee_{i=0}^{n-1}T^{-i}\hat{\alpha})$

$\leq \log N\left(\vee_{i=0}^{n-1}T^{-i}\hat{\alpha}\right)$

$\leq n\log 2 + \log N\left(\vee_{i=0}^{n-1}T^{-i}\mathcal{U}\right)$ (by (14.1)).

Recalling that

$$h(T) \geq h(T,\mathcal{U}) = \lim_{n\to+\infty}\frac{1}{n}H(\vee_{i=0}^{n-1}T^{-i}\mathcal{U})$$

and

$$h_\mu(T,\alpha) = \lim_{n\to+\infty}\frac{1}{n}H_\mu(\vee_{i=0}^{n-1}T^{-i}\alpha)$$

we see that $h_\mu(T,\hat{\alpha}) \leq \log 2 + h(T)$. Moreover, by Corollary 8.6.1 we have that

$$|h_\mu(T,\hat{\alpha}) - h_\mu(T,\alpha)| \leq H_\mu(\alpha|\hat{\alpha}) + H_\mu(\hat{\alpha}|\alpha)$$

$$= -\sum_{C\in\alpha}\sum_{\hat{C}\in\hat{\alpha}}\mu(C\cap\hat{C})\log\left(\frac{\mu(C\cap\hat{C})}{\mu(\hat{C})}\right)$$

$$-\sum_{C\in\alpha}\sum_{\hat{C}\in\hat{\alpha}}\mu(C\cap\hat{C})\log\left(\frac{\mu(C\cap\hat{C})}{\mu(C)}\right) < 1,$$

say, providing ϵ was sufficiently small.

Since α was arbitrary, we see that

$$h_\mu(T) = \sup\{h_\mu(T,\alpha) : \alpha \text{ is a finite partition}\} \leq h(T) + \log 2 + 1.$$

Finally, we can apply the argument to iterates T^k ($k \geq 1$) to see that $h_\mu(T^k) \leq h(T^k) + \log 2 + 1$. By Corollary 3.8.1 we know that $h(T^k) = kh(T)$. The following gives the analogous result for measure theoretic entropy.

SUB-LEMMA 14.1.2 (ABRAMOV'S THEOREM). For $k \geq 1$, $h_\mu(T^k) = kh_\mu(T)$.

PROOF. Given any partition α we observe that

$$h_\mu\left(T^k, \vee_{i=0}^{n-1}T^{-i}\alpha\right) = \lim_{n\to+\infty}\frac{1}{n}H_\mu\left(\vee_{i=0}^{k-1}T^{-ik}\left(\vee_{j=0}^{k-1}T^{-j}\alpha\right)\right)$$

$$= \lim_{N\to+\infty}\frac{k}{N}H_\mu\left(\vee_{i=0}^{N-1}T^{-i}\alpha\right) = kh_\mu(T,\alpha).$$

Given $\epsilon > 0$ we can choose α with $h(T, \alpha) > h_\mu(T) - \epsilon$ so that we have

$$h_\mu(T^k) \geq h_\mu\left(T^k, \vee_{i=0}^{k-1} T^{-i}\alpha\right)$$
$$\geq kh_\mu(T, \alpha) \geq kh_\mu(T) - k\epsilon.$$

Since $\epsilon > 0$ is arbitrary we see that $h_\mu(T^k) \geq kh_\mu(T)$.

To get the reverse inequality, notice that $h_\mu\left(T^k, \alpha\right) \leq h_\mu\left(T^k, \vee_{i=0}^{k-1} T^{-i}\alpha\right)$, using Lemma 8.6. Given $\epsilon > 0$ we can choose α with $h_\mu(T^k, \alpha) > h_\mu(T^k) - \epsilon$ and then

$$kh_\mu(T) \geq kh(T, \alpha) = h_\mu\left(T^k, \vee_{i=0}^{k-1} T^{-i}\alpha\right)$$
$$\geq h\left(T^k, \alpha\right) > h_\mu(T^k) - \epsilon.$$

Since $\epsilon > 0$ is arbitrary we see that $h_\mu(T^k) \leq kh_\mu(T)$. ∎

We can now complete the proof of (1) since

$$h_\mu(T) = \lim_{k \to +\infty} \frac{h_\mu(T^k)}{k}$$
$$\leq \lim_{k \to +\infty} \frac{h(T^k)}{k} + \lim_{k \to +\infty} \frac{\log 2 + 1}{k} = h(T).$$

∎

PROOF OF (2). It suffices to show that given $\delta > 0$ there exists a T-invariant probability measure μ with $h_\mu(T) \geq h(T) - \delta$. We want to choose $\epsilon > 0$ sufficiently small that $\lim_{n \to +\infty} \frac{1}{n} \log(s(n, \epsilon)) \geq h(T) - \delta$, where $s(n, \epsilon)$ is the maximal cardinality of an (n, ϵ)-separating set. We can find a subsequence $n_i \to +\infty$ such that $\frac{1}{n_i} \log(s(n_i, \epsilon)) = h(T)$. Let S_{n_i} be such an (n_i, ϵ)-separated set.

For each n_i we can define a (possibly non-invariant) probability measure

$$\nu_{n_i} = \frac{1}{s(n_i, \epsilon)} \sum_{x \in S_{n_i}} \delta_x.$$

In order to arrive at a T-invariant probability measure we can consider an accumulation point μ (in the weak-star topology) of the measures

$$\mu_{n_i} = \frac{1}{n_i} \sum_{r=0}^{n_i - 1} (T^r)^* \nu_{n_i}.$$

By replacing $\{n_i\}$ by a sub-sequence, if necessary, we can assume that $\mu_{n_i} \to \mu$.

Let want to consider a finite partition $\alpha = \{A_1, \dots, A_k\}$ such that

(1) $\operatorname{diam}(A_i) < \epsilon$, $i = 1, \dots, k$; and
(2) $\mu(\partial A_i) = 0$, for $i = 1, \dots, k$.

Since S_{n_i} is an (n_i, ϵ)-separated set we know that each set $C \in \alpha^{(n_i)} :=$ $\vee_{j=0}^{n_i-1} T^{-j}\alpha$ contains at most one point $x = x_C \in S_{n_i}$. Thus of the sets in S_{n_i} there are $s(n_i, \epsilon)$ sets with ν_{n_i}-measure $\frac{1}{s(n_i, \epsilon)}$ and the remainder have ν_{n_i}-measure zero. In particular, we see that

$$\log(s(n_i, \epsilon)) = - \sum_{C \in \alpha^{(n_i)}} \nu_{n_i}(C) \log \nu_{n_i}(C). \tag{14.2}$$

In order to take limits in a sensible way we fix first $1 < N < n_i$ and then $0 \le j \le N - 1$. We can write

$$\alpha^{(n_i)} = \vee_{i=0}^{n_i-1} T^{-i}\alpha = \left(\vee_{\substack{l=j \ (\text{mod } N) \\ 0 \le l \le n_i-N}} T^{-l} \left(\vee_{i=0}^{N-1} T^{-i}\alpha \right) \right) \vee \left(\vee_{i \in E} T^{-i}\alpha \right)$$

where $E = \{0, 1, \ldots, j-1\} \cup \{M_j, M_j+1, \ldots, n_i-1\}$, with $M_j = N \left[\frac{n_i-j}{N} \right]$, has cardinality at most $2N$.

SUB-LEMMA 14.1.3. *Given measurable partitions β and γ we have that*

$$H_{\nu_{n_i}}(\beta \vee \gamma) \le H_{\nu_{n_i}}(\beta) + H_{\nu_{n_i}}(\gamma)$$

PROOF. For *invariant* measures, this would be an immediate consequence of Lemma 8.4 (and Corollary 8.4.1). However, although in chapter 8 we assumed that the ambient measures were invariant, this property was not used at this stage and the result remains true without it. ∎

In particular, we have that

$$- \sum_{C \in \alpha^{(n_i)}} \nu_{n_i}(C) \log \nu_{n_i}(C)$$

$$\le \sum_{\substack{l=j \ (\text{mod } N) \\ 0 \le l \le N-n_i}} \left(- \sum_{C \in T^{-l}\alpha^{(N)}} \nu_{n_i}(C) \log \nu_{n_i}(C) \right)$$

$$+ \sum_{i \in E} \left(- \sum_{C \in T^{-i}\alpha} \nu_{n_i}(C) \log \nu_{n_i}(C) \right) \tag{14.3}$$

$$\le \sum_{r=0}^{M_j} \left(- \sum_{D \in \alpha^{(N)}} (T^{rN+j})^* \nu_{n_i}(D) \log((T^{rN+j})^* \nu_{n_i}(D)) \right)$$

$$+ 2N \log k$$

(where for $l = rN + j$ there is a natural correspondence between $D \in \alpha^{(N)}$ and $C \in T^{-l}\alpha^{(N)}$ with $(T^l)^*\nu_{n_i}(D) := \nu_{n_i}(T^{-l}D) = \nu_{n_i}(C)$). Summing the inequalities (14.3) over $j = 0, \ldots, N - 1$ we have by (14.2)

$$N \log(s(n_i, \epsilon))$$

$$\leq \sum_{l=0}^{n_i-1} \left(\sum_{D \in \alpha^{(N)}} (T^l)^*\nu_{n_i}(D) \log((T^l)^*\nu_{n_i}(D)) \right) + 2N^2 \log k \qquad (14.4)$$

SUB-LEMMA 14.1.4. *Let α be a measurable partition and let ν_1 and ν_2 be (not necessarily invariant) probabilty measures; then given $0 \leq a \leq 1$ we have that*

$$\sum_{A \in \alpha} [a\nu_1 + (1-a)\nu_2](A) \log[a\nu_1 + (1-a)\nu_2](A)$$

$$\leq a \left(\sum_{A \in \alpha} \nu_1(A) \log\nu_1(A) \right) + (1-a) \left(\sum_{A \in \alpha} \nu_2(A) \log\nu_2(A) \right).$$

PROOF. This follows immediately since $t \mapsto -t \log t$ is convex. ∎

Dividing (14.4) by $n_i N$ we get that

$$\frac{1}{n_i} \log(s(n_i, \epsilon))$$

$$\leq \frac{1}{n_i} \sum_{r=0}^{n_i-1} \left(-\frac{1}{N} \sum_{D \in \alpha^{(N)}} (T^r)^*\nu_{n_i}(C) \log((T^r)^*\nu_{n_i}(C)) \right) + \frac{2N \log k}{n_i}$$

$$\leq - \sum_{C \in \alpha^{(N)}} \mu_{n_i}(C) \log\mu_{n_i}(C) + \frac{2N \log k}{n_i}$$

where we have used Sub-lemma 14.1.4 repeatedly for the last line.

Since we have assumed $\mu(\partial A_i) = 0$, letting $n_i \to +\infty$ (with N fixed) we have that

$$- \sum_{C \in \alpha^{(N)}} \mu_{n_i}(C) \log\mu_{n_i}(C) \to H_\nu(\alpha^{(N)}).$$

This means that

$$h(T) - \delta \leq \lim_{n_i \to +\infty} \frac{1}{n_i} \log(s(n_i, \epsilon))$$

$$\leq \frac{1}{N} H_\nu(\alpha^{(N)}) + \lim_{n_i \to +\infty} \frac{2N^2 \log k}{n_i}$$

$$= \frac{1}{N} H_\nu(\alpha^{(N)}).$$

Letting $N \to +\infty$ we have that

$$h(T) - \delta \leq \lim_{N \to +\infty} \frac{1}{N} H_\nu(\alpha^{(N)}) = h_\mu(\alpha) \leq h_\mu(T).$$

Since $\delta > 0$ is arbitrary this completes the proof. ■

14.3 Comments and reference

The proof we give is due to Misiurewicz [1]. Theorem 14.1 (1) was originally due to Goodman. Theorem 14.1 (2) was subsequently proved by Walters.

References

1. M. Misiurewicz, *A short proof of the variational principle for a* $\mathbb{Z}_+^N$ *action on a compact space*, Astérisque **40** (1976), 147-187.

CHAPTER 15

INVARIANT MEASURES FOR
COMMUTING TRANSFORMATIONS

In this chapter we describe an important conjecture of Furstenberg and related work of Rudolph.

15.1 Furstenberg's conjecture and Rudolph's theorem

Consider the transformations

  (i) $S : \mathbb{R}/\mathbb{Z} \to \mathbb{R}/\mathbb{Z}$ defined by $S(x) = 2x$ (mod 1), and
  (ii) $T : \mathbb{R}/\mathbb{Z} \to \mathbb{R}/\mathbb{Z}$ defined by $T(x) = 3x$ (mod 1).

(For a mnemonic aid: S stands for "second" and T for "third".) It is easy to see that these transformations commute, i.e. $ST = TS$).

Recall that the S-invariant probability measures form a convex weak-star compact set $\mathcal{M}_S$ (and similarly, the T-invariant probability measures form a convex weak-star compact set $\mathcal{M}_T$).

We want to describe the probability measures which are both T-invariant and S-invariant (i.e. the intersection $\mathcal{M}_S \cap \mathcal{M}_T$). We need only consider the (S, T)-ergodic measures μ in $\mathcal{M}_S \cap \mathcal{M}_T$ (i.e. those probability measures invariant under both S and T for which the only Borel sets B with $T^{-n}S^{-m}B = B \ \forall n, m \geq 0$ have either $\mu(B) = 0$ or 1, since these are the extremal measures in $\mathcal{M}_S \cap \mathcal{M}_T$).

FURSTENBERG'S CONJECTURE. *The only (S, T)-ergodic measures are the Haar-Lebesgue measure and measures supported on a finite set.*

Notice that the Haar-Lebesgue measure ν has entropies $\log 2$ and $\log 3$, respectively, for the transformations S and T, and any finitely supported measure always has zero entropy with respect to either S or T. The following partial solution is due to D.J. Rudolph.

THEOREM 15.1 (RUDOLPH). *The only (S, T)-ergodic measure μ which has non-zero entropy (w.r.t. either S or T) is the Haar-Lebesgue measure.*

15.2 The proof of Rudolph's theorem

We begin with a few comments.

  (1) Haar-Lebesgue measure ν on the unit circle is characterized as the only probability measure invariant under all rotations on the circle.

Moreover, it is the only measure invariant under all rotations $x \mapsto x + a(\text{mod } 1)$, where a is any rational number $a = \frac{j}{2^k}$.

(2) For $n \geq 1$ and $f \in L^2(X, \mathcal{B}, \mu)$ we can write that

$$E(f|T^{-n}\mathcal{B})(x) = \sum_{y \in T^{-n}T^n x} \frac{f(y)}{(T^n)'(y)}$$

where $T'(x) = \frac{d\mu T}{d\mu}$ (and similarly for S).

(3) We can write $(S^n)'(x) = S'(S^{n-1}x) \ldots S'(x)$. If we knew that S' is constant (almost everywhere) then by the martingale theorem we would have that

$$\int f(x+a)d\mu = \lim_{n \to +\infty} \int E(f(x+a)|S^{-n}\mathcal{B})$$

$$= \lim_{n \to +\infty} \int E(f(x)|S^{-n}\mathcal{B})$$

$$= \int f(x)d\mu$$

and thus we know that ν is the Haar-Lebesgue measure.

(4) Since $ST = TS$ we have that

$$T'(Sx) \cdot S'(x) = (TS)'(x) = (ST)'(x) = S'(Tx) \cdot S'(x).$$

In particular, we can write $\frac{S'(Tx)}{T'(Sx)} = \frac{S'(x)}{T'(x)}$.

We begin with the following simple (but fundamental) Sub-lemma.

SUB-LEMMA 15.1.1. $S'(Tx) = S'(x)$ for almost all x.

PROOF. We begin by claiming that $E(S'|T^{-1}\mathcal{B})(x) = S'(Tx)$. To see this we observe that

$$E(S'|T^{-1}\mathcal{B})(x) = \sum_{y:Ty=Tx} \frac{S'(y)}{T'(y)}$$

$$= \sum_{y:Ty=Tx} \frac{S'(Ty)}{T'(Sy)} \text{ (by (4) above)}$$

$$= S'(Tx) \left(\sum_{y:Ty=Tx} \frac{1}{T'(Sy)} \right)$$

$$= S'(Tx)$$

where we have used that there is a bijection between $\{y : Ty = Tx\}$ and $\{w : Tw = T(Sx)\}$ to write that

$$\sum_{y:Ty=Tx} \frac{1}{T'(Sy)} = \sum_{w:Tw=T(Sx)} \frac{1}{T'(w)} = 1.$$

This proves the claim. To complete the proof of the Sub-lemma we need only show that $E(S'|T^{-1}\mathcal{B})(x) = S'(x)$. However, since $E(.|T^{-1}\mathcal{B})(x) : L^2(X,\mathcal{B}, \mu) \to L^2(X,\mathcal{B},\mu)$ is a positive operator which is a contraction and

$$||E(S'|T^{-1}\mathcal{B})||_2 = ||S'T|_2| = ||S'||_2,$$

we indeed see that $S'T = E(S'|T^{-1}\mathcal{B})(x) = S'(x)$. ∎

If we knew that T was ergodic we could now deduce S' is constant. Unfortunately, we don't know this and a little more work is required.

DEFINITION. We let $\mathcal{A}_1 \subset \mathcal{B}$ denote the *smallest* sub-sigma-algebra for which $S'(x)$ is measurable.

In the course of the proof we shall establish that $\mathcal{A}_1 \subset S^{-1}\mathcal{B}$ (which, by the definition of $\mathcal{A}_1$, will imply that $S'(x)$ is constant).

Similarly, we can introduce the sub-sigma-algebras $\mathcal{A}_1 \subset \mathcal{A}_2 \subset \ldots \subset \mathcal{A}_n \subset \ldots \subset \mathcal{B}$ where $\mathcal{A}_n$ is the smallest sub sigma-algebra for which all of the functions $S'(x), (S^2)'(x), \ldots ,(S^n)'(x)$ are measurable.

SUB-LEMMA 15.1.2. *For each $n \geq 1$ we have that*

(a) $S^{-1}\mathcal{A}_n \subset \mathcal{A}_{n+1}$,
(b) $T^{-1}\mathcal{A}_n = \mathcal{A}_n$.

PROOF.

(a) If we write $(S^{n+1})'(x) = (S^n)'(Sx) \cdot S'(x)$ then, since by hypothesis S^n and S' are $\mathcal{A}_n$-measurable, the right hand side is measurable with respect to $S^{-1}\mathcal{A}_n$.
(b) Since $S'(Tx) = S'(x)$ we also see that

$$(S^n)'(Tx) = S'(Tx) \cdot S'(STx) \ldots S'(S^{n-1}Tx)$$
$$= S'(x) \cdot S'(Sx) \ldots S'(S^{n-1}x)$$
$$= (S^n)'(x).$$

But the right hand side is $\mathcal{A}_n$-measurable by hypothesis. ∎

DEFINITION. We write $\mathcal{A} = \vee_{n=1}^{\infty} \mathcal{A}_n$. The above lemma guarantees that $T^{-1}\mathcal{A} = \mathcal{A}$ and $S^{-1}\mathcal{A} \subset \mathcal{A}$.

We now move on to entropy considerations. Let $\gamma = \{[0, \frac{1}{6}], [\frac{1}{6}, \frac{2}{6}], .., [\frac{5}{6}, 1]\}$ denote the partition into intervals of length one sixth.

SUB-LEMMA 15.1.3. *There exists a sequence $s_n \to +\infty$ such that the partitions*

$$\vee_{i=0}^{s_n} S^{-i}\gamma = \left\{ \left[0, \frac{1}{6 \cdot 2^{s_n}}\right], \left[\frac{1}{6 \cdot 2^{s_n}}, \frac{2}{6 \cdot 2^{s_n}}\right], \ldots, \left[\frac{6 \cdot 2^{s_n} - 1}{6 \cdot 2^{s_n}}, 1\right] \right\} \quad and$$

$$\vee_{i=0}^{n} T^{-i}\gamma = \left\{ \left[0, \frac{1}{6 \cdot 3^n}\right], \left[\frac{1}{6 \cdot 3^n}, \frac{2}{6 \cdot 3^n}\right], \ldots, \left[\frac{6 \cdot 3^n - 1}{6 \cdot 3^n}, 1\right] \right\}$$

have the property that every element of either partition is contained in at most four elements of the other partition.

PROOF. For each $n \geq 1$ we choose the values $s_n \geq 1$ such that $3^{n-1} \leq 2^{s_n} \leq 3^n$. The lengths of the intervals for each partition are $\frac{1}{6 \cdot 2^{s_n}}$ and $\frac{1}{6 \cdot 3^n}$ and thus their ratios are bounded above and below by 3 and $\frac{1}{3}$, respectively. This is enough to complete the proof. ∎

In what follows we shall make frequent use of the basic identity for entropy: $H(\alpha \vee \beta | \mathcal{C}) = H(\alpha | \mathcal{C} \vee \beta) + H(\beta | \mathcal{C})$.

Recall that the *entropies* of the transformations are given by

$$h(T) := \lim_{n \to +\infty} \frac{1}{n} H(\vee_{i=0}^{n-1} T^{-i}\gamma)$$

and

$$h(S) := \lim_{k \to +\infty} \frac{1}{k} H(\vee_{i=0}^{k-1} S^{-i}\gamma).$$

The following sub-lemma shows similar limits involving the sigma-algebra $\mathcal{A}$.

SUB-LEMMA 15.1.4. *The following limits exist and are equal to the entropies:*

$$h(T) = \lim_{n \to +\infty} \frac{1}{n} H(\vee_{i=0}^{n-1} T^{-i}\gamma | \mathcal{A})$$

and

$$h(S) = \lim_{n \to +\infty} \frac{1}{s_n} H(\vee_{i=0}^{s_n-1} S^{-i}\gamma | \mathcal{A}).$$

PROOF. We begin with an argument which is borrowed from the standard entropy identities. We see that for any $n, m \geq 0$ we have that

$$H(\vee_{i=0}^{n+m-1} T^{-i}\gamma | \mathcal{A})$$
$$= H(\vee_{i=0}^{n-1} T^{-i}\gamma | \mathcal{A}) + H(\vee_{i=n}^{n+m-1} T^{-i}\gamma | \mathcal{A} \vee (\vee_{i=0}^{n-1} T^{-i}\gamma))$$
$$\leq H(\vee_{i=0}^{n-1} T^{-i}\gamma | \mathcal{A}) + H(\vee_{i=n}^{n+m-1} T^{-i}\gamma | \mathcal{A})$$
$$= H(\vee_{i=0}^{n-1} T^{-i}\gamma | \mathcal{A}) + H(\vee_{i=0}^{m-1} T^{-i}\gamma | \mathcal{A})$$

(where for the last equality we use that $T^{-1}\mathcal{A} = \mathcal{A}$). Thus by subadditivity the limit $h(T|\mathcal{A}) := \lim_{n \to +\infty} \frac{1}{n} H(\vee_{i=0}^{n-1} T^{-i}\gamma | \mathcal{A})$ exists. By the basic equalities for entropy we see that

$$H(\vee_{i=0}^{s_n-1} S^{-i}\gamma | \mathcal{A}) - H(\vee_{i=0}^{n-1} T^{-i}\gamma | \mathcal{A})$$
$$= H\left(\left(\vee_{i=0}^{s_n} S^{-i}\gamma\right) \vee \left(\vee_{i=0}^{n-1} T^{-i}\gamma\right) | \mathcal{A}\right) + H\left(\vee_{i=0}^{s_n} S^{-i}\gamma | \left(\vee_{i=0}^{n-1} T^{-i}\gamma\right) \vee \mathcal{A}\right)$$
$$- H\left(\left(\vee_{i=0}^{s_n} S^{-i}\gamma\right) \vee \left(\vee_{i=0}^{n-1} T^{-i}\gamma\right) | \mathcal{A}\right) - H\left(\vee_{i=0}^{n-1} T^{-i}\gamma | \left(\vee_{i=0}^{s_n} S^{-i}\gamma\right) \vee \mathcal{A}\right)$$

and so we can identify the limit as

$$h(T|\mathcal{A}) = \lim_{n \to +\infty} \frac{1}{n} H(\vee_{i=0}^{n-1} T^{-i}\gamma | \mathcal{A})$$
$$= \lim_{n \to +\infty} \frac{1}{n} \left(H\left(\vee_{i=0}^{n-2} T^{-i}\gamma | \mathcal{A}\right) + H\left(T^{-(n-1)}\gamma | \mathcal{A} \vee \left(\vee_{i=0}^{n-2} T^{-i}\gamma\right)\right)\right)$$
$$= h(T)$$

since

$$h(T) = \lim_{n \to +\infty} \frac{1}{n} \left(H(\vee_{i=0}^{n-2} T^{-i}\gamma)\right)$$

and

$$H\left(T^{-(n-1)}\gamma | \mathcal{A} \vee \left(\vee_{i=0}^{n-2} T^{-i}\gamma\right)\right) \leq H\left(T^{-(n-1)}\gamma | \mathcal{A}\right) = H(\gamma | \mathcal{A}) < +\infty.$$

By sublemma 15.1.3 we have that the final expression above is bounded (independently of n) and thus we have that the following limit exists

$$h(S|\mathcal{A}) := \lim_{n \to +\infty} \frac{1}{n} H(\vee_{i=0}^{s_n-1} T^{-i}\gamma | \mathcal{A}).$$

Moreover, this argument gives that $h(T|\mathcal{A}) = \frac{\log 3}{\log 2} h(S|\mathcal{A})$.

Observe that if we replace $\mathcal{A}$ by the trivial sigma-algebra then the same argument gives that $h(T) = \frac{\log 3}{\log 2} h(S)$. Comparing these identities we see that $h(S) = h(S|\mathcal{A})$. ∎

We now apply Sub-lemma 15.1.4 to show that $\mathcal{A} \subset \vee_{i=1}^{\infty} S^{-i}\gamma$, which is essentially the end of the proof.

SUB-LEMMA 15.1.5. $H(\mathcal{A}| \vee_{i=1}^{\infty} S^{-1}\mathcal{B}) = 0.$

PROOF. By the basic equality for entropy we have that

$$
\begin{aligned}
H(\mathcal{A}| \vee_{i=1}^{\infty} S^{-i}\gamma) &= H(\gamma \vee \mathcal{A}| \vee_{i=1}^{\infty} S^{-i}\beta) - H(\gamma| \vee_{i=1}^{\infty} S^{-i}\gamma \vee \mathcal{A}) \\
&= H(\gamma| \vee_{i=1}^{\infty} S^{-1}\beta) - H(\gamma| \vee_{i=1}^{\infty} S^{-i}\gamma \vee \mathcal{A}) \qquad (15.1) \\
&= h(S) - H(\gamma| \vee_{i=1}^{\infty} S^{-i}\gamma \vee \mathcal{A})
\end{aligned}
$$

(where we have used that $H(\gamma \vee \mathcal{A}| \vee_{i=1}^{\infty} S^{-i}\beta) = H(\gamma| \vee_{i=1}^{\infty} S^{-i}\beta) = h(S)$).
We next observe that

$$
\begin{aligned}
&H(\vee_{i=0}^{n-1} S^{-i}\gamma|\mathcal{A}) \\
&= H(\gamma| \vee_{i=1}^{n-1} S^{-i}\gamma \vee \mathcal{A}) + H(\vee_{i=1}^{n-1}S^{-i}\gamma|\mathcal{A}) \\
&\leq H(\gamma| \vee_{i=1}^{n-1} S^{-i}\gamma \vee \mathcal{A}) + H(\vee_{i=0}^{n-2}S^{-i}\gamma|\mathcal{A}) \\
&\cdots \\
&\leq H(\gamma| \vee_{i=1}^{n-1} S^{-i}\gamma \vee \mathcal{A}) + H(\gamma| \vee_{i=1}^{n-2} S^{-i}\gamma \vee \mathcal{A}) + \ldots + H(\gamma|\mathcal{A})
\end{aligned}
$$

(This argument is a modification of the standard entropy proof that $h(S) = H(\gamma| \vee_{i=1}^{\infty} S^{-i}\gamma)$.) Thus from the definition of $h(S|\mathcal{A})$ we have that

$$
\begin{aligned}
h(S|\mathcal{A}) &:= \lim_{n \to +\infty} \frac{1}{n} H(\vee_{i=0}^{n-1}S^{-i}\gamma|\mathcal{A}) \\
&= \lim_{n \to +\infty} \frac{1}{n} H(\gamma| \vee_{i=1}^{n-1} S^{-i}\gamma \vee \mathcal{A}) + \lim_{n \to +\infty} \frac{1}{n} H(\vee_{i=1}^{n-1}S^{-i}\gamma|\mathcal{A}) \\
&= H(\gamma| \vee_{i=1}^{\infty} S^{-i}\gamma).
\end{aligned}
$$

$$(15.2)$$

Comparing (15.1) and (15.2) we see that

$$
0 \leq H(\gamma| \vee_{i=1}^{\infty} S^{-i}\gamma) \leq h(T) - h(T|\mathcal{A}) = 0. \qquad \blacksquare
$$

To finish off the proof of Theorem 15.1 we need only recall that $H(\mathcal{A}| \vee_{i=1}^{\infty} S^{-i}\gamma) = 0$ implies that $\mathcal{A} \subset \vee_{i=1}^{\infty}S^{-i}\gamma$.

Repeating the argument with S replaced by S^k for $k = 1, 2, \ldots$ we see that $\mathcal{A} \subset \cap_{n=0}^{\infty}S^{-n}\mathcal{B}$. In particular, this shows that $(S^n)'(y)$ is constant for $y \in \{w|S^n x = S^n w\}$ (almost everywhere).

We observe that since $h(S) > 0$ (equivalently $h(T) > 0$), there must be a set of positive measure on which S has two pre-images (otherwise S would be invertible almost everywhere and then have entropy zero). Moreover we claim that the set with two S pre-images is invariant under S and T. By *ergodicity* of (S, T) we see that almost all points have two pre-images.

This suffices to apply the argument in comment (3). $\blacksquare$

15.3 Comments and references

The original proof of Rudolph had a symbolic formulation [2]. The proof we give here is a version due to Parry [1].

References

1. W. Parry, *Squaring and cubing the circle*, Ergodic Theory, Proceedings of the Warwick Symposium on $\mathbb{Z}^d$-actions (M. Pollicott and K. Schmidt, ed.), C.U.P., Cambridge, 1996, pp. 177-183.
2. D. Rudolph, ×2 *and* ×3 *invariant measures and entropy*, Ergod. Th. and Dynam. Sys. **10** (1990), 395-406.

CHAPTER 16

MULTIPLE RECURRENCE
AND SZEMEREDI'S THEOREM

In this chapter we shall present a famous application of ergodic theory to number theory. The theorem we shall prove is an improvement on Van der Waerden's theorem (Theorem 2.1) originally proved by Szemeredi. The proof we give is due to Furstenberg.

16.1 Szemeredi's theorem on arithmetic progressions

To state the theorem we begin with the following definition.

DEFINITION. We say that a set of integers $\mathcal{N} \subset \mathbb{Z}$ has *positive density* if

$$\delta(\mathcal{N}) := \limsup_{N \to +\infty} \frac{1}{2N+1} \mathrm{Card}\{-N \leq n \leq N : n \in \mathcal{N}\} > 0.$$

(This quantity is called the *(upper) density* of $\mathcal{N}$.)

The following result generalizes the result of Van der Waerden in chapter 2.

THEOREM 16.1 (SZEMEREDI). *If $\mathcal{N} \subset \mathbb{Z}$ has positive density then for all $k \geq 1$ there exists an arithmetic progression of length k in $\mathcal{N}$ (i.e. $\forall k \geq 1$, $\exists a, b \in \mathbb{Z}$, $b \neq 0$ such that $a + ib \in \mathcal{N}$, $i = 0, \ldots, k-1$).*

REMARKS.

(i) It is clear from the definition of δ that in the special case $\mathcal{N} = \mathbb{Z}$ we have that $\delta(\mathbb{Z}) = 1$. Moreover, if we consider a finite partition of the integers $\mathbb{Z} = \mathcal{N}_1 \cup \ldots \mathcal{N}_k$ then $\sup_{1 \leq i \leq k} \delta(N_i) \geq \frac{1}{k}$, and thus at least one of the elements in this partition has positive density. Thus Van der Waerden's theorem can be considered as a corollary to Szemeredi's theorem.

(ii) Finite sets obviously have density zero. The set of odd (or even) numbers has density $\frac{1}{2}$.

(iii) The prime numbers have zero density (i.e. $\delta(\{\mathrm{primes}\}) = 0$) and so Szemeredi's theorem does not apply. It is an open problem as to whether the conclusion holds for this set.

A historical comment. The above result was conjectured by Erdös and Turan in 1936. For the special case of $k = 3$ the above theorem was proved by Roth in 1952. The special case $k = 4$ was then proved by Szemeredi in 1969 before proving the general theorem in 1975.

In 1977 Furstenberg gave an alternative proof using ergodic theory.

16.2 An ergodic proof of Szemeredi's theorem

The key to the ergodic theory proof of Szemeredi's theorem is the following multi-dimensional version of the Poincaré recurrence theorem.

THEOREM 16.2. *Let $T : (X, \mathcal{B}) \to (X, \mathcal{B})$ be a measurable transformation and let μ be a T-invariant probability measure. For any $A \in \mathcal{B}$ with $\mu(A) > 0$ and $k \geq 1$ there exists $n \geq 1$ such that*

$$\mu\left(A \cap T^{-n}A \cap T^{-2n}A \cap \ldots \cap T^{-kn}A\right) > 0. \qquad (MR)$$

NOTATION. We can say that the set A satisfies the *multiple recurrence property* when the condition (MR) holds. A stronger condition, which we refer to as *uniform multiple recurrence*, is when for all $k \geq 1$ the following condition holds:

$$\liminf_{N \geq 1} \frac{1}{N} \sum_{n=1}^{N} \mu\left(A \cap T^{-n}A \cap T^{-2n}A \cap \ldots \cap T^{-kn}A\right) > 0. \qquad (UMR)$$

PROOF OF THEOREM 16.1 ASSUMING THEOREM 16.2. Let $X = \prod_{n=-\infty}^{\infty} \{0, 1\}$ and let $\sigma : X \to X$ be the associated shift map (i.e. $(\sigma x)_n = x_{n-1}$). Given $\mathcal{N} \subset \mathbb{N}$ we can define a sequence $x = (x_n)_{n \in \mathbb{Z}} \in X$ by

$$x_n = \begin{cases} 1 \text{ if } n \in \mathcal{N}, \\ 0 \text{ if } n \notin \mathcal{N}. \end{cases}$$

For any $n \geq 1$ we can define a probability measure μ_n on X by

$$\mu_n = \frac{1}{2n+1} \sum_{i=-n}^{i=n} \delta_{\sigma^i x}.$$

Since the space of probability measures on X is compact (in the weak-star topology) we can choose a limit point μ of $\{\mu_n : n \geq 1\}$. This will have the following properties:

(a) μ is a σ-invariant probability measure (since for any continuous function $f : X \to \mathbb{R}$ we have that

$$\int f\sigma d\mu = \lim_{n \to +\infty} \frac{1}{2n+1} \sum_{i=-n+1}^{i=n+1} f(\sigma^i x)$$

$$= \lim_{n \to +\infty} \frac{1}{2n+1} \sum_{i=-n}^{i=n} f(\sigma^i x) = \int f d\mu);$$

(b) If we choose $A = \{y = (y_n)_{n \in \mathbb{Z}} : y_0 = 1\}$ then $\mu(A) > 0$ (since $\mu(A) = \limsup_{n \to +\infty} \mu_n(A) = \limsup_{n \to +\infty} \frac{1}{2n+1}(\mathrm{Card}\{\mathcal{N} \cap [-n, n]\}) = \delta(\mathcal{N}) > 0)$.

We can now apply Theorem 16.2 to $\sigma : X \to X$, the measure μ and the set A, to deduce that $\forall k \geq 1, \exists n \geq 1$ with

$$\mu\left(A \cap \sigma^{-n}A \cap \sigma^{-2n}A \cap \ldots \cap \sigma^{-kn}A\right) > 0.$$

From the construction of μ we see that for (open and closed) sets $A \cap \sigma^{-n}A \cap \sigma^{-2n}A \cap \ldots \cap \sigma^{-kn}A$ there are values of $N \geq 1$ such that $\mu_N(A \cap \sigma^{-n}A \cap \sigma^{-2n}A \cap \ldots \cap \sigma^{-kn}A) > 0$. In particular, for some $-N \leq i \leq N$ we have that $\sigma^i x \in A \cap \sigma^{-n}A \cap \sigma^{-2n}A \cap \ldots \cap \sigma^{-kn}A$.

Finally, from the definition of x this means that the arithmetic progression $i + jk \in \mathcal{N}$, for $j = 0, \ldots, n$.

∎

16.3 The proof of Theorem 16.2

In this section we present an overview of the proof of Theorem 16.2. Details of the technical propositions used are presented in the appendix to this section.

16.3.1: (UMR) for weak-mixing systems, weak-mixing extensions and compact systems.
The (UMR) condition is easier to establish if we assume that $T : (X, \mathcal{B}) \to (X, \mathcal{B})$ under additional assumptions. For example:

PROPOSITION 16.3. *Let* $T : (X, \mathcal{B}) \to (X, \mathcal{B})$ *be weak-mixing; then* T *satisfies (UMR).*

To introduce the definition of a weak-mixing extension we first recall a useful fact on skew products (which can be found in the work of Rohlin [4]).

LEMMA (ROHLIN). *Let* $T : (X, \mathcal{B}) \to (X, \mathcal{B})$ *be a measurable map, and let* μ *be a T-invariant ergodic measure. Given a T-invariant sub-sigma-algebra* $\mathcal{A} \subset \mathcal{B}$ *(i.e. if $A \in \mathcal{A}$ then $T^{-1}A \in \mathcal{A}$) there exists a skew product*

$$\begin{cases} S : X_1 \times X_2 \to X_1 \times X_2, \\ S(x_1, x_2) = (S_1(x_1), S_2(x_1, x_2)) \end{cases}$$

(with respect to measure spaces $(X_1, \mathcal{B}_1, \mu_1)$ and $(X_2, \mathcal{B}_2, \mu_2)$) such that

(i) *there exists an isomorphism* $\psi : (X_1 \times X_2, \mathcal{B}_1 \times \mathcal{B}_2, \mu_1 \times \mu_2) \to (X, \mathcal{B}, \mu)$ *between S and T,*

(ii) *the images $\psi(A \times X_2)$, $A \in \mathcal{B}_1$, correspond to sets in $\mathcal{A} \subset \mathcal{B}$,*

(iii) *$S_2(x_1, \cdot) : X_2 \to X_2$ preserves μ_2 (for a.e. (μ_1) x_1).*

The property of sub-sigma-algebras we want to specify is closely related to the definition of weak-mixing (cf. chapter 11) and makes use of the above lemma of Rohlin.

We can identify the quadruple $\mathcal{A} = (X, \mathcal{A}, \mu, T)$ with the quadruple $(X_1, \mathcal{B}_1, \mu_1, S_1)$.

DEFINITION. Let $T : (X, \mathcal{B}) \to (X, \mathcal{B})$ be a measurable map and let μ be an ergodic T-invariant probability measure. Given a T-invariant sub-sigma-algebra $\mathcal{A} \subset \mathcal{B}$ we define the $\mathcal{A}$-*cartesian product* to be the skew product $T \times_{\mathcal{A}} T := S_1 \times (S_2 \times S_2)$, i.e.

$$\begin{cases} S_1 \times (S_2 \times S_2) : X_1 \times (X_2 \times X_2) \to X_1 \times (X_2 \times X_2), \\ S_1 \times (S_2 \times S_2)(x_1, x_2, y_2) = (S_1(x_1), S_2(x_1, x_2), S_2(x_1, y_2)). \end{cases}$$

with the product sigma-algebra $\mathcal{B}_1 \times \mathcal{B}_2 \times \mathcal{B}_2$ and product measure $\mu_1 \times \mu_2 \times \mu_2$.

We say that $T : (X, \mathcal{B}, \mu) \to (X, \mathcal{B}, \mu)$ is *weak-mixing (relative to $\mathcal{A}$)* if the $\mathcal{A}$-cartesian product $T \times_{\mathcal{A}} T$ is ergodic.

REMARK. If we take $\mathcal{A}$ to be the trivial sigma-algebra $\{X, \emptyset\}$ then in the Rohlin lemma X_1 is trivial, $X_2 = X$ and $T \times_{\mathcal{A}} T = T \times T$. We see by Proposition 11.8 that T being weak-mixing relative to $\{X, \emptyset\}$ is equivalent to $T : X \to X$ being weak-mixing.

The next result shows that the (UMR) property is preserved under weak-mixing extensions.

PROPOSITION 16.4. *Let $T : (X, \mathcal{B}, \mu) \to (X, \mathcal{B}, \mu)$ be ergodic. Let $T : (X, \mathcal{B}) \to (X, \mathcal{B})$ be a weak-mixing extension of $T : (X, \mathcal{A}) \to (X, \mathcal{A})$ which satisfies (UMR); then $T : (X, \mathcal{B}) \to (X, \mathcal{B})$ also satisfies (UMR).*

Notice that when $\mathcal{A} = \{X, \emptyset\}$ then Proposition 16.4 reduces to Proposition 16.3.

We now introduce a property that complements that of weak mixing.

DEFINITION. The system $(X, \mathcal{B})$ is *compact* if for every $f \in L^2(X, \mu)$ the closure in $L^2(X, \mu)$ of the orbit $\{f \circ T^n\}_{n \geq 0}$ is compact.

EXAMPLE. Let $T : \mathbb{R}/\mathbb{Z} \to \mathbb{R}/\mathbb{Z}$ be a rotation $T(x) = x + \alpha \pmod 1$ $(\alpha \in \mathbb{R})$; then this is a compact system. (The same applies for any rotation on a compact group.)

The next proposition complements Proposition 16.3 by giving a second special case in which (UMR) can be readily established.

PROPOSITION 16.5. *If the system $T : (X, \mathcal{B}) \to (X, \mathcal{B})$ is compact then property (UMR) holds.*

16.3.2: The non-weak-mixing case. In view of Proposition 16.3 we have to concentrate on the alternative where $T : (X, \mathcal{B}) \to (X, \mathcal{B})$ is not weak-mixing.

PROPOSITION 16.6. *If the system $T : (X, \mathcal{B}) \to (X, \mathcal{B})$ is not weak-mixing then there exists a non-trivial compact factor $T : (X, \mathcal{B}_1) \to (X, \mathcal{B}_1)$ which satisfies property (UMR).*

DEFINITION. We denote by $\mathcal{F}$ the family of factors $\mathcal{A} \subset \mathcal{B}$ for which $T : (X, \mathcal{A}) \to (X, \mathcal{A})$ satisfies property (UMR).

By Proposition 16.6, $\mathcal{F}$ is non-empty (and non-trivial). The main result on $\mathcal{F}$ is the following.

PROPOSITION 16.7. *There exists a maximal factor in $\mathcal{F}$ (i.e. $\exists \mathcal{B}_\infty \in \mathcal{F}$ such that $\mathcal{B}_1 \subset \mathcal{B}_\infty, \forall \mathcal{B}_1 \in \mathcal{F}$).*

16.3.3: (UMR) for compact extensions. Given $\mathcal{A} \subset \mathcal{B}$, a T-invariant sigma-algebra, and thus by Rohlin's lemma a quadruple $(X_1, \mathcal{B}_1, \mu_1, S_1)$ identified with $(X, \mathcal{A}, \mu, T)$, and $x_1 \in X$ we write $\mu_{x_1}(B) = \mathrm{E}(\chi_B | \mathcal{A})(x_1)$. We need the following definition.

DEFINITION. A function $f \in L^2(X, \mathcal{B}, \mu)$ is *almost periodic* (AP) relative to the factor $\mathcal{A} \subset \mathcal{B}$ if for every $\delta > 0$ there exist functions $F_1, \dots, F_n \in L^2(X, \mathcal{A}, \mu)$ such that for every $j \in \mathbb{N}$ we have $\inf_{1 \le s \le n} \|f \circ T^j - F_s\|_{L^2(\mu_y)} < \delta$ for almost all $x_1 \in X$.

We let $\mathcal{P}(\mathcal{A})$ denote the set of all almost periodic functions in $L^2(X, \mu)$, relative to $\mathcal{A}$.

DEFINITION. If $\mathcal{P}(\mathcal{A}) \subset L^2(X, \mathcal{B}, \mu)$ is dense then $T : (X, \mathcal{B}) \to (X, \mathcal{B})$ is a *compact extension* of $T : (X, \mathcal{A}) \to (X, \mathcal{A})$

The following result complements Proposition 16.4 by showing that (UMR) is also preserved under compact extensions.

PROPOSITION 16.8. *Let $T : (X, \mathcal{C}) \to (X, \mathcal{C})$ be a compact extension of $T : (X, \mathcal{A}) \to (X, \mathcal{A})$ which satisfies (UMR); then $T : (X, \mathcal{C}) \to (X, \mathcal{C})$ also satisfies (UMR).*

16.3.4: The last step. The final ingredient in the proof of Theorem 16.2 is the following.

PROPOSITION 16.9. *Let $T : (X, \mathcal{B}) \to (X, \mathcal{B})$ be a not relatively weak-mixing extension of $T : (X, \mathcal{A}) \to (X, \mathcal{A})$; then there exists an intermediate factor $T : (X, \mathcal{C}) \to (X, \mathcal{C})$ which is a non-trivial compact extension of $T : (X, \mathcal{A}) \to (X, \mathcal{A})$.*

To finish the proof of Theorem 16.2 we proceed as follows. If $T : (X, \mathcal{B}) \to (X, \mathcal{B})$ is weak-mixing, then we are done (by Proposition 16.3). Alternatively, if $T : (X, \mathcal{B}) \to (X, \mathcal{B})$ is not weak-mixing then there are a further two possibilities: either the maximal sigma-algebra $\mathcal{B}_\infty$ (given by Proposition 16.7) is equal to $\mathcal{B}$ or it is a non-trivial sub-sigma-algebra of $\mathcal{B}$. In the first case the proof is complete because $\mathcal{B} = \mathcal{B}_\infty \in \mathcal{F}$ satisfies (UMR) by definition. In the second case, we may assume that the maximal factor $\mathcal{B}_\infty \subset \mathcal{B}$ is not relatively weak-mixing since otherwise we immediately have that $T : (X, \mathcal{B}) \to (X, \mathcal{B})$ satisfies (UMR), by Proposition 16.4. Under this assumption there exists (by Proposition 16.9) an intermediate non-trivial extension $T : (X, \mathcal{C}) \to (X, \mathcal{C})$ (i.e. $\mathcal{B}_\infty \subset \mathcal{C} \subset \mathcal{B}$) which is relatively compact (see diagram).

$$\overbrace{\mathcal{B}_\infty \;\subset\; \mathcal{C}}^{\text{compact}} \;\underbrace{\subset\; \mathcal{B}}_{\text{not relatively weak-mixing}}$$

But then (by Proposition 16.8) we know that $T : (X, \mathcal{C}) \to (X, \mathcal{C})$ satisfies property (UMR). Since by construction $\mathcal{C} \neq \mathcal{B}_\infty$ this gives a contradiction to $\mathcal{B}_\infty$ being maximal, and so the second case cannot occur. Thus the proof of Theorem 16.2 is complete. ∎

16.4 Appendix to section 16.3

In this rather lengthy appendix, we shall give the omitted proofs of propositions from section 16.3.

16.4.1 The proofs of Propositions 16.3 and 16.4. Proposition 16.3 is a special case of Proposition 16.4. To simplify notation, we shall write $E(f|\mathcal{A})$ for $E(f|\mathcal{A})\psi^{-1}$.

LEMMA 16.10. *Let $T : (X, \mathcal{B}) \to (X, \mathcal{B})$ be a relatively weak-mixing extension of $T : (X, \mathcal{A}) \to (X, \mathcal{A})$ and $f, g \in L^\infty(\mu)$. Then*

$$\lim_{N \to \infty} \frac{1}{N} \sum_{n=1}^{N} \int \left\{ E(f(gT^n)|\mathcal{A}) - E(f|\mathcal{A})E(gT^n|\mathcal{A}) \right\}^2 d\mu_1 = 0.$$

In particular, for the characteristic functions χ_A and χ_B we have

$$\int \frac{1}{N} \sum_{n=1}^{N} |\mu_{x_1}(A \cap T^{-n}B) - \mu_{x_1}(A)\mu_{x_1}(B)| d\mu_1(x_1) \to 0, \quad \text{as } n \to \infty.$$

PROOF. We can assume that $E(f|\mathcal{A}) = 0$ (otherwise we simply replace f by $f - E(f|\mathcal{A})$).

Define $f \otimes_A f : X_1 \times (X_2 \times X_2) \to X_1 \times (X_2 \times X_2)$ by $f \otimes_A f(x_1, x_2, y_2) = f(x_1, x_2) f(x_1, y_2)$; then

$$
\lim_{N \to +\infty} \frac{1}{N} \sum_{n=1}^{N} \int \{E(f.g \circ T^n | \mathcal{A})\}^2 \, d\mu_1
$$

$$
= \lim_{N \to +\infty} \int (f \otimes_A f) \cdot \left(\frac{1}{N} \sum_{n=1}^{N} g \otimes_A g (T \times_A T)^n \right) d(\mu_1 \times \mu_2 \times \mu_2). \tag{16.1}
$$

By assumption $T \times_A T$ is ergodic and so by the Birkhoff ergodic theorem

$$
\lim_{n \to +\infty} \frac{1}{N} \sum_{n=1}^{N} g \otimes_A g (T \times_A T)^n = \int g \otimes_A g \, d(\mu_1 \times \mu_2 \times \mu_2)
$$

$$
= \int \{E(g|\mathcal{A})\}^2 \, d\mu_1
$$

(where the last line follows from the definitions of $g \otimes_A g$).

Hence the limit in (16.1) is

$$
\int (f \otimes_A f) \, d(\mu_1 \times \mu_2 \times \mu_2) \cdot \left\{ \int \{E(g|\mathcal{A})\}^2 \, d\mu_1 \right\}^2 = 0
$$

since $\int (f \otimes_A f) \, d(\mu_1 \times \mu_2 \times \mu_2) = \int \{E(f|\mathcal{A})\}^2 \, d\mu_1 = 0$ (because we assumed $E(f|\mathcal{A}) = 0$).
∎

LEMMA 16.11. *Let $T : (X, \beta) \to (X, \beta)$ be a relatively weak-mixing extension of $\mathcal{A}$. Then $T \times_A T$ is also relatively weak-mixing extension of $\mathcal{A}$.*

PROOF. This is an immediate consequence of Lemma 16.10 and the equality $E(gf|\mathcal{A}) = gE(f|\mathcal{A})$, whenever $g \in L^1(X, \mathcal{A})$.
∎

LEMMA 16.12. *Let $T : (X, C) \to (X, C)$ be a relatively weak-mixing extension of $T : (X, \mathcal{A}) \to (X, \mathcal{A})$. Then if $f_l \in L^\infty(\mu), l = 1, \ldots, k$, we have,*

$(1)_k \lim_{N \to \infty} \frac{1}{N} \sum_{n=1}^{N} \int \left\{ E(\prod_{l=0}^{k} f_l T^{ln} | \mathcal{A}) - \prod_{l=0}^{k} E(f_l T^{ln} | \mathcal{A}) \right\}^2 d\mu_1 = 0$,

$(2)_k \lim_{N \to \infty} \| \frac{1}{N} \sum_{n=1}^{N} \left(\prod_{l=1}^{k} f_l T^{ln} - \prod_{l=1}^{k} E(f_l | \mathcal{A}) T^{ln} \right) \|_{L^2(\mu)} = 0$.

PROOF. First we show $(2)_k$ by induction. For $k = 1$ the limit $(2)_1$ follows from the ergodic theorem since

$$
\begin{cases}
\frac{1}{N} \sum_{n=1}^{N} f_1(T^n x) \to \int f_1 d\mu & \text{(in L^2 norm)}, \\
\frac{1}{N} \sum_{n=1}^{N} E(f_1|\mathcal{A})(T^n x) \to \int E(f_1|\mathcal{A}) d\mu = \int f_1 d\mu & \text{(in L^2 norm)}.
\end{cases}
$$

Assume for the inductive step that the result has been established for $k-1$ functions (i.e. $(2)_{k-1}$ is valid).

By the simple identity

$$\prod_{l=1}^{k} a_l - \prod_{l=1}^{k} b_l = \sum_{j=1}^{k} \left(\prod_{l=1}^{j-1} a_l \right) (a_j - b_j) \left(\prod_{l=j+1}^{k} b_l \right),$$

with $a_1, \ldots, a_k, b_1, \ldots, b_k \in \mathbb{R}$ we can write

$$(\int |\frac{1}{N} \sum_{n=1}^{N} (f_1(T^n x) f_2(T^{2n} x) \ldots f_k(T^{kn} x)$$

$$- E(f_1|\mathcal{A})(T^n x) E(f_2|\mathcal{A})(T^{2n} x) \ldots E(f_k|\mathcal{A})(T^{kn} x))|^2 d\mu(x))^{\frac{1}{2}}$$

$$\leq \sum_{j=1}^{k} (\int |\frac{1}{N} \sum_{n=1}^{N} (E(f_1|\mathcal{A})(T^n x) \ldots E(f_{j-1}|\mathcal{A})(T^{(j-1)n} x)$$

$$\times f_j(T^{jn} x) f_{j+1}(T^{(j+1)n} x) \ldots f_k(T^{kn} x))|^2 d\mu(x))^{\frac{1}{2}}.$$

We fix a choice of $1 \leq j \leq k$ and we may assume without loss of generality that $E(f_j|\mathcal{A}) = 0$ (otherwise we need only replace f_j by $f_j - E(f_j|\mathcal{A})$). Thus it suffices to show that

$$\alpha_j(N) := \int |\frac{1}{N} \sum_{n=1}^{N} \beta_j(n)|^2 d\mu(x) \to 0$$

as $N \to +\infty$, where

$$\beta_j(n) = E(f_1|\mathcal{A})(T^n x) \ldots E(f_{j-1}|\mathcal{A})(T^{(j-1)n} x) f_j(T^{jn} x) \ldots f_k(T^{kn} x).$$

For any $1 \leq m \leq N$ we can now bound

$$\alpha_j(N)$$

$$\leq \int |\frac{1}{N} \sum_{n=1}^{N} (\frac{1}{m} \sum_{i=0}^{m-1} \beta_j(n+i)|^2 d\mu(x) + \frac{2m|f_1|_\infty \ldots |f_k|_\infty}{N}$$

$$\leq \frac{1}{N} \sum_{n=1}^{N} \left(\int |\frac{1}{m} \sum_{i=0}^{m-1} \beta_j(n+i)|^2 d\mu(x) \right) + \frac{2m|f_1|_\infty \ldots |f_k|_\infty}{N}$$

$$= \frac{1}{N} \sum_{n=1}^{N} \left(\frac{1}{m^2} \sum_{i,i'=0}^{m-1} \int \beta_j(n+i)\beta_j(n+i') d\mu(x) \right) + \frac{2m|f_1|_\infty \ldots |f_k|_\infty}{N}.$$

Collecting together similar terms we get that

$$\alpha_j(N) \leq \frac{1}{N} \sum_{n=1}^{N} \sum_{r=-m}^{m} \frac{(m-|r|)}{m^2} \int \prod_{s=1}^{j-1} \left(E(f_s|\mathcal{A}) \cdot E(f_s|\mathcal{A}) \circ T^{sr}\right) (T^{(s-1)n}x)$$

$$\times \prod_{t=j}^{k} \left(f_t \cdot f_t \circ T^{tr}\right) (T^{(t-1)n}x) d\mu(x)$$

$$+ \frac{2m|f_1|_\infty \cdots |f_k|_\infty}{N}.$$

We can now write that

$$\limsup_{N \to +\infty} |\alpha_j(N)|^2 \leq \sum_{r=-m}^{m} \frac{m-|r|}{m^2} ||E(f_1|\mathcal{A})||_\infty^2 \cdots ||E(f_{j-1}|\mathcal{A})||_\infty^2$$

$$\times ||E(f_{j+1}|\mathcal{A})||_\infty^2 \cdots ||E(f_k|\mathcal{A})||_\infty^2$$

$$\times |\int E(f_j \circ T^{jr} . f_j|\mathcal{A}) d\mu|.$$

since in the limit we can replace the terms $\prod_{t=j+1}^{k} \left(f_t \cdot f_t \circ T^{tr}\right) (T^{(t-1)n}x)$ by the terms $\prod_{t=j+1}^{k} \left(E(f_t|\mathcal{A}) \cdot E(f_t|\mathcal{A}) \circ T^{tr}\right) (T^{(t-1)n}x)$ using the inductive hypothesis.

Finally, we know that the averages over terms $\int E(f_j|\mathcal{A}) \circ T^{jr} E(f_j|\mathcal{A}) d\mu$, for $-m \leq r \leq m$, are small for large m (since using Lemma 16.10 we can show that the terms tend to 0 for a sub-sequence of density one). This completes the proof of $(2)_k$.

To prove $(1)_k$ we proceed as follows. By $(2)_k$ we know that for $f \in L^\infty(\mu)$,

$$\lim_{N \to +\infty} \left\{ \int f_0 \left(\frac{1}{N} \sum_{n=1}^{N} \prod_{l=1}^{k} f_l \circ T^{ln} \right) d\mu \right\}$$

$$= \lim_{N \to +\infty} \left\{ \int f_0 \left(\frac{1}{N} \sum_{n=1}^{N} \prod_{l=1}^{k} E(f_l|\mathcal{A}) \circ T^{ln} \right) d\mu \right\}.$$

Since $T \times_\mathcal{A} T$ is also relatively weak-mixing (by Lemma 16.11) we can replace f_l by $f_l \otimes_\mathcal{A} f_l$ and T by $T \times_\mathcal{A} T$ so that we obtain

$$\lim_{N \to +\infty} \left\{ \int f_0 \otimes_\mathcal{A} f_0 \left(\frac{1}{N} \sum_{n=1}^{N} \prod_{l=1}^{k} (f_l \otimes_\mathcal{A} f_l) \circ (T \times_\mathcal{A} T)^{ln} \right) d(\mu_1 \times \mu_2 \times \mu_2) \right\}$$

$$= \lim_{N \to +\infty} \left\{ \int f_0 \otimes_\mathcal{A} f_0 \left(\frac{1}{N} \sum_{n=1}^{N} \prod_{l=1}^{k} E(f_l \otimes_\mathcal{A} f_l|\mathcal{A})) \circ (T \times_\mathcal{A} T)^{ln} \right) d(\mu_1 \times \mu_2 \right.$$

$$\left. \times \mu_2 \right).$$

$$(16.2)$$

A similar argument to the proof of Lemma 16.10 allows us to see that the right hand side of (16.2) is equal to

$$\int \left\{ E(\prod_{l=1}^{k} f_k | \mathcal{A}) \right\}^2 d\mu_1 \cdot \int \left\{ E(f_0|\mathcal{A}) \right\}^2 d\mu_1. \tag{16.3}$$

Let us assume for the present that $E(f_0|\mathcal{A}) = 0$, then we see that (16.3) (and thus the right hand side of (16.2)) is identically zero. On the other hand, the left hand side of (16.2) is equal to

$$\lim_{N\to+\infty} \frac{1}{N} \sum_{n=1}^{N} \int \left\{ E\left(f_0 \times \prod_{l=1}^{k} f_l \circ T^{ln} \Big| \mathcal{A} \right) \right\}^2 d\mu_1$$

$$= \lim_{N\to+\infty} \frac{1}{N} \sum_{n=1}^{N} \int \left\{ E\left(\prod_{l=0}^{k} f_l \circ T^{ln} \Big| \mathcal{A} \right) \right\}^2 d\mu_1$$

showing $(1)_k$.

It remains to consider the case that $E(f_0|\mathcal{A}) \neq 0$. We then write $f_0 = (f_0 - E(f_0|\mathcal{A})) + E(f_0|\mathcal{A})$. The proof of $(1)_k$ for $(f_0 - E(f_0|\mathcal{A}))$ is as above and for the proof of $(1)_k$ for $E(f_0|\mathcal{A})$ we may write

$$\lim_{N\to\infty} \frac{1}{N} \sum_{n=1}^{N} \int \left\{ E\left(E(f_0|\mathcal{A}) \prod_{l=1}^{k} f_l \circ T^{ln} \Big| \mathcal{A} \right) - \prod_{l=0}^{k} E(f_l \circ T^{ln} | \mathcal{A}) \right\}^2 d\mu_1$$

$$= \lim_{N\to\infty} \frac{1}{N} \sum_{n=1}^{N} \int E(f_0|\mathcal{A})^2 \left\{ E(\prod_{l=1}^{k} f_l \circ T^{ln} | \mathcal{A}) - \prod_{l=1}^{k} E(f_l \circ T^{ln} | \mathcal{A}) \right\}^2 d\mu_1$$

$$\leq \| E(f_0|\mathcal{A})^2 \|_\infty$$

$$\times \lim_{N\to\infty} \frac{1}{N} \sum_{n=1}^{N} \int \left\{ E(\prod_{l=1}^{k} f_l \circ T^{ln} | \mathcal{A}) - \prod_{l=1}^{k} E(f_l \circ T^{ln} | \mathcal{A}) \right\}^2 d\mu_1 = 0$$

using $(1)_{k-1}$.

■

To complete the proof of Proposition 16.4 we proceed as follows. Let $A \in \mathcal{B}$ with $\mu(A) > 0$. Let $\epsilon > 0$ be a small number so that for $A_1 = \{x : E(\chi_A|\mathcal{B}_1) \geq \epsilon\}$ we have $\mu_1(A_1) > 0$. It follows from Lemma 16.12 and $E(\chi_A|\mathcal{B}_1) \geq \epsilon\chi_{A_1}$ that we have the following:

$$\frac{1}{N} \sum_{j=1}^{N} \mu(\cap_{l=0}^{k} T^{-lj}A) > \frac{1}{2}\epsilon^{k+1} \frac{1}{N} \sum_{j=1}^{N} \mu_1(\cap_{l=0}^{k} S_1^{-lj}A_1),$$

for all $k > 0$.

■

16.4.2 The proof of Proposition 16.5. Let f satisfy $0 \leq f \leq 1$. We first note that if $F_0, \ldots, F_k$ are measurable functions with $0 \leq F_i \leq 1$ satisfying $||f - F_i||_2 < \epsilon$ $(i = 0, \ldots, k)$, then

$$\left| \int \prod_{l=0}^{k} F_l d\mu - \int f^{k+1} d\mu \right| \leq \sum_{l=0}^{k} \int \prod_{j=0}^{l-1} F_j |F_l - f| f^{k-l} d\mu$$

$$\leq \epsilon \sum_{l=0}^{k} \int \prod_{j=0}^{l-1} F_j f^{k-l} d\mu$$

$$\leq (k+1)\epsilon.$$

So if we put $a = \int f^{k+1} d\mu > 0$ and choose $\epsilon < a/(k+1)$, we have $\int \prod_{l=0}^{k} F_l d\mu \geq (k+1)\epsilon - a > 0$.

LEMMA 16.13. *For a set of n of positive lower density, $||f \circ T^n \cdot f - f||_2 < \frac{\epsilon}{k}$ for $l = 1, \ldots, k$.*

PROOF. Since $\mathrm{cl}\{f \circ T^n\} \subset L^2(\mu)$ is compact we can find a finite set $\{f \circ T^{m_1}, f \circ T^{m_2}, \ldots, f \circ T^{m_r}\}$ which is $\frac{\epsilon}{k}$-separated, i.e.

$$\left(\int |f \circ T^{m_i} - f \circ T^{m_j}|^2 d\mu \right)^{\frac{1}{2}} \geq \frac{\epsilon}{k}, \quad \text{for } 1 \leq i < j \leq r.$$

We can assume that r is the maximal cardinality of all such subsets.

For all $n \geq 0$ we have that $\{f \circ T^{n+m_1}, f \circ T^{n+m_2}, \ldots, f \circ T^{n+m_r}\}$ is again $\frac{\epsilon}{k}$-separated. For each n there exists $1 \leq i(n) \leq r$ such that $\left(\int |f - f \circ T^{n+m_{i(n)}}| d\mu \right)^{\frac{1}{2}} \geq \frac{\epsilon}{k}$. In particular, the sequence $\{n + m_{i(n)}\}_{n=0}^{\infty}$ is a sequence of positive lower density with the required property. ∎

16.4.3 The proof of Proposition 16.6.

DEFINITION. A function $f \in L^2(\mu)$ is *almost periodic* if its orbit closure is compact.

In particular, $T : (X, \mathcal{B}) \to (X, \mathcal{B})$ is compact if every $f \in L^2(X, \mathcal{B})$ is almost periodic.

LEMMA 16.14. *If $T \times T : (X \times X, \mathcal{B} \times \mathcal{B}) \to (X \times X, \mathcal{B} \times \mathcal{B})$ is not ergodic, then there exists a non-constant function $f \in L^2(\mu)$ which is almost periodic.*

PROOF. Since $T \times T$ is not ergodic we can choose $g(x, y)$ to be a non-constant $(T \times T)$-invariant function in $L^2(\mu \times \mu)$. In the case that T is ergodic, the existence of a non-constant almost periodic function is trivial, so we suppose that T is not ergodic. Define a metric d on X by

$$d(x, y) = \int |g(x, z) - g(y, z)| d\mu(z).$$

If we identify points that are 0 distance, then we have an invariant sub-sigma-algebra (i.e., factor) so that T acts on this factor space as an isometry. Furthermore if we can show that the metric space is totally bounded (i.e. for any $\epsilon > 0$ there exist a finite number $N(\epsilon)$ of points which are ϵ-separated) then we can apply the well-known result: *for any function f defined on a totally bounded metric space the orbit closure of $\{f \circ T^n\}$ is compact when T is an isometry.* Without loss of generality, we can assume that the T-invariant function $\int g(x,y)d\mu(x)$, which is constant by ergodicity, vanishes (else we subtract the constant from g). Since g is not identically 0 there exists a function $h \in L^2(\mu)$ satisfying $\int g(x,y)h(y)d\mu(y) \neq 0$ on a set of positive measure. Now define a function $H(x) = \int g(x,y)h(y)d\mu(y)$ which is non-constant. Then we see that the function H is the desired almost periodic function. In fact, $H \circ T^n(x) = \int g(x,y)h \circ T^n(y)d\mu(y)$ (because of the T-invariance of μ) and the integral operator

$$\begin{cases} \mathcal{G} : L^2(\mu) \to L^2(\mu), \\ \mathcal{G}\phi(x) = \int g(x,y)\phi(y)d\mu(y) \end{cases}$$

is a compact operator. Since the closure of $\{H \circ T^n\}_{n>0}$ coincides with the closure of $\{\mathcal{G}(h \circ T^n)\}_{n>0}$ and the norms of $h \circ T^n$ are constant, we have the desired result. ∎

Now we suppose that $T : (X, \mathcal{B}) \to (X, \mathcal{B})$ is not weak-mixing. By lemma 16.14 we have a non-constant function $f \in L^2(X, \mathcal{B})$ which is almost periodic. Recall that a subset of a complete metric space has compact closure if and only if for any $\epsilon > 0$ there are finitely many balls of radius less than $\epsilon > 0$ which are cover the subset. It follows from this fact that the set of $f \in L^2(X, \mathcal{B})$ which are almost periodic is a closed linear subspace of $L^2(X, \mathcal{B})$. In fact, we can see that the set of almost periodic functions is closed under the (lattice) operations $(h_1, h_2) \mapsto \max(h_1, h_2)$ and $(h_1, h_2) \mapsto \min(h_1, h_2)$. Let $\mathcal{B}_0$ be the smallest sigma-algebra of sets with respect to which f is measurable. In particular, every characteristic function χ_A $(A \in \mathcal{B}_0)$ is also almost periodic. Let $\mathcal{B}_1$ be the smallest sigma-algebra of sets with respect to which $f, f \circ T, \dots, f \circ T^n, \dots$ are measurable. Since f is almost periodic if and only if $f \circ T$ is almost periodic we know that each χ_A, $A \in \mathcal{B}_1$, is almost periodic. If for every $A \in \mathcal{B}_1$ we know $\chi_A \in \mathcal{P}(\mathcal{B}_1)$ then any $f' \in L^2(X, \mathcal{B}_1)$ must be almost periodic because the set of almost periodic functions is a closed linear subspace of $L^2(X, \mathcal{B})$ as we have mentioned above. Finally, we have a non-trivial T-invariant sigma-algebra $\mathcal{B}_1$ so that the factor $T : (X, \mathcal{B}_1) \to (X, \mathcal{B}_1)$ is compact.

Using Proposition 16.5 the proof of Proposition 16.6 is complete. ∎

16.4.4 The proof of Proposition 16.7.

We wish to apply Zorn's lemma. Thus it suffices to show that every chain contains a maximal element, i.e. if

$\{A_\alpha\} \subset \mathcal{F}$ is a totally ordered chain then $\cup_\alpha A_\alpha \in \mathcal{F}$ (where this represents the sigma-algebra generated by the union).

Let $A \in \cup_\alpha A_\alpha$ with $\mu(A) > 0$ and fix $k > 0$. Take $\rho = \frac{1}{2(k+1)}$; then for α_0 sufficiently large we can choose $A'_0 \in \mathcal{A}_{\alpha_0}$ with

$$\mu(A \triangle A'_0) < \frac{1}{4}\rho\mu(A). \tag{16.4}$$

By assumption $T : (X, \mathcal{A}_{\alpha_0}) \to (X, \mathcal{A}_{\alpha_0})$ satisfies (UMR). We can use Rohlin's lemma to find a system $S_1 : (X_1, \mathcal{B}_1, \mu_1) \to (X_1, \mathcal{B}_1, \mu_1)$ and a map $\pi : X \to X_1$ so that $\mathcal{A}_{\alpha_0} = \pi^{-1}(\mathcal{B}_1)$. Let $A''_0 = \pi(A'_0)$; then by (16.3) we have $\mu(A'_0) \geq \mu(A) - \frac{1}{4}\rho\mu(A)$. This implies that $\mu_1(x_1 \in A''_0 : \mu_{x_1}(A) < 1 - \eta) < \frac{1}{4}\mu(A)$.

Write $A_0 = \{x_1 \in A''_0 : \mu_{x_1}(A) > 1 - \eta\}$. Then $A_0 \in \mathcal{B}_1$ and

$$\mu_1(A_0) > \mu_1(A''_0) - \frac{1}{4}\mu(A) = \mu(A'_0) - \frac{1}{4}\mu(A) > \frac{1}{2}\mu(A).$$

By hypothesis $T : (X, \mathcal{A}_{\alpha_0}) \to (X, \mathcal{A}_{\alpha_0})$ satisfies (UMR). We claim that for every $j > 0$

$$\frac{1}{2}\mu_1\left(A_0 \cap S_1^{-j}A_0 \cap \ldots \cap S_1^{-kj}A_0\right) < \mu\left(A \cap T^{-j}A \cap \ldots \cap T^{-kj}A\right) \tag{16.5}$$

To see this it suffices to show that for $x_1 \in A_0 \cap S_1^{-j}A_0 \cap \ldots \cap S_1^{-kj}A_0$

$$\mu_{x_1}\left(A \cap T^{-j}A \cap \ldots \cap T^{-kj}A\right) > \frac{1}{2} \tag{16.6}$$

because (16.5) follows from (16.6) by integration. But if $x_1 \in S_1^{-lj}A_0$, $l = 0, \ldots, k$, then by definition of A_0 we see that $\mu_{x_1}(S_1^{-lj}A_0) > 1 - \eta$ and (16.6) follows easily.

Averaging over $j = 1, \ldots, N$ and letting $N \to +\infty$ we get

$$\liminf_{N \to +\infty} \frac{1}{N} \sum_{j=1}^{N} \mu\left(A \cap T^{-j}A \cap \ldots \cap T^{-kj}A\right) > 0.$$

∎

16.4.5 Proof of Proposition 16.8. We shall show that for $A \in \mathcal{C}$ with $\mu(A) > 0$ and for $k > 0$ we have

$$\liminf_{N \to +\infty} \frac{1}{N} \sum_{j=1}^{N} \mu\left(\cap_{l=1}^{k} T^{-jl}A\right) > 0. \tag{16.7}$$

Since the above inequality follows for any subset of $\mathcal{A}$, we can assume (without loss of generality) that $\mu_{x_1}(A) \geq \frac{1}{2}\mu(A)$ for $x_1 \in A_1$ with $\mu_1(A_1) > \frac{1}{2}\mu(A)$ and $\mu_{x_1}(A) = 0$ for $x_1 \notin A_1$ (by removing from A those fibres projecting to x_1 satisfying $\mu_{x_1}(A) \leq \frac{1}{2}\mu(A)$). The claim follows.

LEMMA 16.15. χ_A *is almost periodic*

PROOF. Compactness of $(X, \mathcal{C})$ implies that for every $\epsilon > 0$ there exists an almost periodic function f' such that $\left(\int |f - f'|^2 d\mu \right)^{\frac{1}{2}} < \epsilon^2$. This means that the set $E_\epsilon = \{x_1 \in X_1 : \int |f - f'|^2 \, d\mu_{x_1}\} \geq \epsilon^2\}$ satisfies $\mu_1(E_\epsilon) < \epsilon^2$. (If this were not the case we would have a contradiction since

$$
\epsilon^4 > \int \int |f - f'| d\mu_{x_1} d\mu_1(x_1)
$$

$$
\geq \int_{E_\epsilon} \int |f - f'|^2 d\mu_{x_1} d\mu_1(x_1)
$$

$$
\geq \epsilon^2 \mu_1(E_\epsilon) \geq \epsilon^4 .)
$$

Let $A_\epsilon = A \cap E_\epsilon^c$ then on every fibre and for every $j > 0$ either

(1) $\int |\chi_{A_\epsilon} \circ T^j - f' \circ T^j|^2 d\mu_{x_1} < \epsilon^2$, or

(2) $\int |\chi_{A_\epsilon} \circ T^j| d\mu_{x_1} = 0$.

Since f' is almost periodic we have from the definition that for each $\delta > 0$ there exist $F_1, \dots, F_n \in L^2(\mu)$ such that $\inf_{0 \leq s \leq n} \left(\int |\chi_{A_\epsilon} \circ T^j - F_s|^2 d\mu_{x_1} \right)^{\frac{1}{2}} < \delta + \epsilon$, for almost all x_1 and $j > 0$.

We can replace A_ϵ by the intersection in $\mathcal{A}$; then providing we choose a sequence $\{\epsilon_j\}_{j>0}$ with $\sum_{j>0} \epsilon_j^2 < +\infty$ the above procedure gives a sequence of sets A_{ϵ_j}. The set $A_\infty = \cap_{j>0} A_{\epsilon_j}$ has non-zero measure and the characteristic function χ_A is almost periodic. ∎

By the above lemma $f = \chi_A$ is almost periodic. Given x_1 let $\oplus_{l=0}^k L^2(\mu_{x_1})$ have the norm

$$
\|(f_1, \dots, f_k)\|_{x_1} = \max_{1 \leq i \leq k} \left(\int |f_i|^2 d\mu_{x_1} \right)^{\frac{1}{2}}.
$$

Let $V(k, f, x_1)$ be the set of vectors of the form $(f, f \circ T^n, \dots, f \circ T^{kn})_y$, $n \in \mathbb{N}$. This has compact closure in $L^2(\mu_{x_1})$.

Denote by $V^*(k, f, x_1)$ the subset of $V(k, f, x_1)$ consisting of elements all of whose components are non-zero ($\|\| \geq \frac{1}{2} (\mu(A))^{\frac{1}{2}}$).

Given $x_1 \in A_1$ and $\epsilon > 0$ we denote by $M(\epsilon, x_1)$ the maximum cardinality of an ϵ-separated set in $V^*(k, f, x_1)$. Since $V^*(k, f, x_1)$ has compact closure, $M(\epsilon, x_1)$ is bounded for $x_1 \in A_1$.

We choose $\frac{\mu(A)}{10k} > \epsilon_0 > 0$, $\eta > 0$ and $A_2 \subset A_1$ with $\mu_1(A_2) > 0$ such that $M(\epsilon, x_1) = M$, $\forall \epsilon_0 - \eta \leq \epsilon \leq \epsilon_0$, $\forall x_1 \in A_2$. Since $M(\epsilon, x_1)$ is an integer valued monotone decreasing function of ϵ it is locally constant (except at a countable set of ϵ) and is a measurable function.

Choose $x_1^0 \in A_2$ and $m_1, \dots, m_M$ for x_1^0 such that $\{(f, f \circ T^{m_j}, \dots, f \circ T^{km_j})_{x_1^0} : j = 1, \dots, M\}$ is a maximal ϵ_0-separated set in $V^*(k, f, x_1^0)$.

As a function on the factor space X_1

$$x_1 \mapsto \left(\int |f \circ T^{lm_i} - f \circ T^{lm_j}| d\mu_{x_1} \right)^{\frac{1}{2}}$$

is measurable (for $1 \leq i < j \leq M$ and for $l = 0, \ldots, k$).

Without loss of generality we can assume that every neighbourhood of the image of x_1^0 has positive μ-measure in A_2. Now we let $A_3 \subset A_2$, $\mu(A_3) > 0$, be the set such that

$$\left(\int |f \circ T^{lm_i} - f \circ T^{lm_j}| d\mu_{x_1^0} \right)^{\frac{1}{2}} - \left(\int |f \circ T^{lm_i} - f \circ T^{lm_j}| d\mu_{x_1} \right)^{\frac{1}{2}} < \eta$$

for $y \in A_3$.

Assume that $T : (X, \mathcal{A}) \to (X, \mathcal{A})$ satisfies (UMR). Let $n > 0$ satisfy $\mu_1 \left(\cap_{l=0}^{k} S_1^{-nl} A_3 \right) > 0$ and fix $x_1 \in \cap_{l=0}^{k} S_1^{-nl} A_3$. It follows from the definition of A_3 and $V^*(k, f, x_1)$ that

$$A_3 \subset \cap_{l=0}^{k} \left(S_1^{-lm_j} A_1 \cap S^{-ln} A_1 \right)$$

for $j = 1, \ldots, M$. Since the vectors $\{(f, f \circ T^{n+m_j}, \ldots, f \circ T^{k(n+m_j)})_{x_1}\}_{j=1}^{M}$ are $(\epsilon_0 - \eta)$-separated in $V^*(k, f, x_1)$, these form a maximal set which is $(\epsilon_0 - \eta)$-dense in $V^*(k, f, x_1)$.

Applying the argument in the proof of Lemma 16.13 to $(f, \ldots, f) \in V^*(k, f, x_1)$ we can choose $j > 0$ such that $(f, f \circ T^{n+m_j}, \ldots, f \circ T^{k(n+m_j)})_{x_1}$ is ϵ_0-close to it. Then

$$\mu_y \left(\cap_{l=0}^{k} T^{-l(n+m_j)} A \right) = \int \prod_{l=0}^{k} f \circ T^{n+m_j} d\mu_y \geq \int \chi_A d\mu_{x_1} - k\epsilon_0 > C\mu(A)$$

for some $C > 0$. Summing over $j = 1, \ldots, N$, integrating over $x_1 \in \cap_{l=0}^{k} S_1^{-ln} A_3$; then averaging over $1 \leq n \leq N$ and letting $N \to +\infty$ finally gives (16.2). ∎

16.4.6 Proof of Proposition 16.9. We can assume $T : (X, \mathcal{B}) \to (X, \mathcal{B})$ is ergodic and use Rohlin's lemma so that we identify $(X, \mathcal{B})$ and $(X_1 \times X_2, \mathcal{B}_1 \times \mathcal{B}_2)$.

Assume that $T \times_{\mathcal{A}} T$ is not ergodic; then there exits a bounded function $g(x, y)$ on $X \times_{\mathcal{A}} X := X_1 \times X_2 \times X_2$ $(= \cup_{x_1 \in X_1} \pi^{-1}(x_1) \times \pi^{-1}(x_1))$ which is invariant under $T \times_{\mathcal{A}} T$ but is not purely a function of $x \in \pi^{-1}(x_1)$, nor purely a function of $y \in \pi^{-1}(x_1)$. We write $g(x, y) = g(x_1, x_2, x_2')$ and define an integral operator

$$\begin{cases} \mathcal{G} : L^2(X, \mathcal{B}) \to L^2(X, \mathcal{B}), \\ \mathcal{G}\phi(x_1, x_2) := \int g(x_1, x_2, x_2') \phi(x_1, x_2') d\mu_2(x_2') \end{cases}$$

which is a compact operator.

By analogy with the proof of Lemma 16.14 we have the following. There exists a function $h \in L^2(X, \mathcal{B}, \mu)$ such that

$$\mathcal{G}h(x_1, x_2) := \int g(x_1, x_2, x_2')h(x_1, x_2')d\mu_2(x_2')$$

is not a function of x_1 alone and $\mathrm{cl}\{(\mathcal{G}h)T^j : j \geq 0\} = \mathrm{cl}\{\mathcal{G}(h \circ T^j) : j \geq 0\} \subset L^2(X, \mathcal{B})$. For $\delta > 0$ there exists $M = M(x_1, \delta) > 0$ such that $\{(\mathcal{G}h) \circ T^j)\}_{j=-M}^M$ is ϵ-dense in $\{(\mathcal{G}h) \circ T^j\}_{j \in \mathbb{N}}$ (in the $L^2(\mu_{x_1})$-norm).

For every $\epsilon > 0$ we choose a sufficiently large $M_{\epsilon,\delta}$ and a set $E(\delta, \epsilon) \subset X_1$ with $\mu_1(E(\delta, \epsilon)) < \epsilon$ such that $M(x_1, \delta) < M_{\epsilon,\delta}$ for all $x_1 \in E(\delta, \epsilon)^c$. For a positive sequence $\{\delta_j\}_{j \in \mathbb{N}}$ (with $\delta_j \to 0$ as $j \to +\infty$) and a positive sequence $\{\epsilon_j\}_{j \in \mathbb{N}}$ (with $\sum_{j=1}^\infty \epsilon_j$ sufficiently small) we define $f : L^2(X, \mathcal{B}) \to L^2(X, \mathcal{B})$:

$$f(x_1, x_2) = \begin{cases} 0 & \text{if } x_1 \in \cup_{j \in \mathbb{N}} E(\delta_j, \epsilon_j), \\ \mathcal{G}h & \text{otherwise.} \end{cases}$$

For each x_1 the integral operator is compact and so $\|f - \mathcal{G}h\|_2 \leq \|g\|_{L^\infty} \cdot \|h\|_{L^2(X)} \sum_{j \in \mathbb{N}} \epsilon_j$. Furthermore, for every $\delta > 0$ and large M the family $\{0\} \cup \{(\mathcal{G}h) \circ T^j\}_{j=-M}^M$ is δ-dense in $\{fT^j\}_{j \in \mathbb{N}}$ (in the $L^2(\mu_{x_1})$-norm for every x_1).

Let $\mathcal{F}$ be the algebra spanned by $\{\mathcal{G}h : g \in L^\infty(X \times_{\mathcal{A}} X), g(T \times_{\mathcal{A}} T) = g, h \in L^\infty(X)\}$; then $\mathcal{F}$ is T-invariant and the almost periodic functions in $\mathcal{F}$ are dense in $\mathcal{F}$.

Let $\mathcal{B}^* \subset \mathcal{B}$ denote the smallest sigma algebra with respect to all of the elements of $\mathcal{F}$ are measurable. $\mathcal{B}^*$ is T-invariant and $\mathcal{B}_1 \subset \mathcal{B}^*$. Moreover, $\mathcal{F} \subset L^2(X, \mathcal{B}^*, \mu)$ is dense and so the set of almost periodic functions is dense in $L^2(X, \mathcal{B}^*, \mu)$

Finally, the desired compact factor corresponds to $T : (X, \mathcal{B}^*) \to (X, \mathcal{B}^*)$. ∎

16.5 Comments and references

Most of the details are taken from the article of Furstenberg, Katznelson and Ornstein [2]. Alternative accounts of the proof appear in [1] and [5] and [3] contains an overview of the proof.

References

1. H. Furstenberg, *Recurrence in Ergodic Theory and Combinatorial Number Theory*, P.U.P., Princeton, N.J., 1981.
2. H. Furstenberg, I. Katznelson, and D. Ornstein, *The ergodic theoretical proof of Szemerédi's theorem*, Bull. Amer. Math. Soc. **7** (1982), 527-552.
3. K. Petersen, *Ergodic Theory*, C.U.P., Cambridge, 1983.
4. V. Rohlin, *Selected topics from the metric theory of dynamical systems*, Transl. Amer. Math. Soc. (2) **49** (1966), 171-209.
5. J.-P. Thouvenot, *La démonstration de Furstenberg du théorème de Szemerédi sur les progressions arithmétiques*, Séminaire Bourbaki, 30 (1977-78), 518.